REVISE AQA GCSE
Science
Further Additional Science
REVISION GUIDE

Series Consultant: Harry Smith   Authors: Sue Kearsey, Nigel Saunders and Peter Ellis

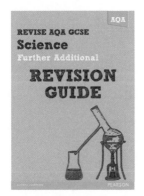

THE REVISE AQA SERIES
Available in print or online

Online editions for all titles in the Revise AQA series are available Spring 2013.

Presented on our ActiveLearn platform, you can view the full book and customise it by adding notes, comments and weblinks.

Print editions

Further Additional Science Revision Guide	9781447942498
Further Additional Science Revision Workbook	9781447942160

Online editions

Further Additional Science Revision Guide	9781447942276
Further Additional Science Revision Workbook	9781447942269

Print and online editions are also available for Science A (Foundation and Higher) and Additional Science (Foundation and Higher).

This Revision Guide is designed to complement your classroom and home learning, and to help prepare you for the exam. It does not include all the content and skills needed for the complete course. It is designed to work in combination with Pearson's main AQA GCSE Science 2011 Series.

To find out more visit:
www.pearsonschools.co.uk/aqagcsesciencerevision

ALWAYS LEARNING   PEARSON

Contents

1-to-1 page match with the **Further Additional Higher** Workbook ISBN 978-1-442-94216-0

- -

A small bit of small print

AQA publishes Sample Assessment Material and the Specification on its website. This is the official content and this book should be used in conjunction with it. The questions in *Now try* this have been written to help you practise every topic in the book. Remember: the real exam questions may not look like this.

Target grades

Target grade ranges are quoted in this book for some of the questions. Students targeting this grade range should be aiming to get most of the marks available. Students targeting a higher grade should be aiming to get all of the marks available.

Into and out of cells

Dissolved substances (SOLUTES) move into and out of cells by DIFFUSION and ACTIVE TRANSPORT.

Diffusion

high concentration of dissolved molecules (concentrated solution)

partially permeable membrane

There is net movement down the concentration gradient.

low concentration of dissolved molecules (dilute solution)

Remember: diffusion is the net movement of molecules from their region of higher concentration to their region of lower concentration. Net movement is the average of the movement of all molecules.

Active transport

partially permeable membrane

Active transport needs energy from respiration.

high concentration of dissolved molecules (concentrated solution)

low concentration of dissolved molecules (dilute solution)

There is net movement against the concentration gradient.

Active transport makes it possible for cells to absorb ions from very dilute solutions, e.g. root cells absorb minerals from soil water, small intestine cells absorb glucose from digested food in the gut into the body.

Worked example   D-C

The diagram shows the results of an experiment. At the start, the level of solution inside the tube and the level of water in the beaker were the same. Name the process that caused the change and explain what happened.   (3 marks)

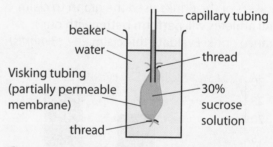

capillary tubing
beaker
water
thread
Visking tubing (partially permeable membrane)
30% sucrose solution
thread

Osmosis is the name given to a special case of diffusion. Osmosis is the net movement of water molecules across a partially permeable membrane.

The process is called osmosis. To start with there was a higher concentration of water molecules in the beaker than in the solution. Water molecules moved into the tubing through the partially permeable membrane. There was a net movement of water into the tubing and the level of the solution rose.

Now try this

1 What is **osmosis**?   (2 marks)
2 (a) Give **one** similarity between diffusion and osmosis.   (1 mark)
   (b) Give **one** difference between diffusion and active transport.   (1 mark)

3 A plant root is treated with a poison that prevents respiration. Explain whether the root cells will still be able to absorb water and mineral ions from a dilute solution.   (4 marks)

Sports drinks

All soft drinks are a solution of substances dissolved in water.

Sports drinks are soft drinks that are designed for use during physical activity.

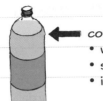

contents:
- water
- sugar
- ions (e.g. sodium, chloride)

during exercise →

Sweating removes water and ions from the body.

Sugars are broken down in respiration to release extra energy for muscles.

→ If the water and ion balance in the body is not kept steady, the cells will not work so well.

→ Sports drinks can replace lost sugars, water and ions.

Some sports drinks are **isotonic** to body fluids. This means they contain the same concentrations of sugars and ions as normal body fluids.

Worked example

target B–A*

This graph shows the length of time that 10 athletes were able to continue vigorous activity while replacing lost fluids with different drinks. The makers of the drinks used the graph to claim that 'All athletes will perform better with our endurance drink.' Evaluate this claim. *(3 marks)*

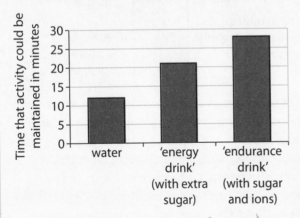

Time that activity could be maintained in minutes

water | 'energy drink' (with extra sugar) | 'endurance drink' (with sugar and ions)

The graph shows that when the athletes drank the endurance drink they exercised vigorously for more than twice as long as when they only drank water, and about a third longer than when they drank the energy drink. This suggests that the endurance drink is better for long periods of vigorous activity, but you can't tell from this if it is better for other kinds of activity. More tests would be needed to check this.

EXAM ALERT!

Questions like this often ask you if a claim made by someone is backed up by data. Make sure that you use the data to answer the question.

Students have struggled with exam questions similar to this – **be prepared!**

Now try this

target G–E

1 How are water and ions lost from the body during activity? *(1 mark)*

target D–C

2 Explain why a sugar-containing energy drink may help an athlete to exercise for longer. *(2 marks)*

target B–A*

3 Some sports drinks are hypotonic (contain a lower concentration of ions than body fluids). Explain why an isotonic sports drink may be better when running a marathon than a hypotonic sports drink. *(3 marks)*

Exchanging materials

As organisms increase in size and complexity, it becomes more difficult for them to exchange gases and solutes with the environment. Special organs are adapted to make exchange efficient. For example, the lungs are adapted to exchange gases, and the small intestine is adapted to exchange solutes.

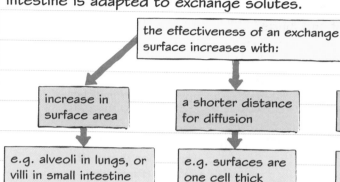

the effectiveness of an exchange surface increases with:

- increase in surface area
  - e.g. alveoli in lungs, or villi in small intestine
- a shorter distance for diffusion
  - e.g. surfaces are one cell thick
- maintenance of a high concentration gradient
  - e.g. animals have an efficient blood supply, lungs are ventilated

The fast removal of substances maintains a steep concentration gradient so diffusion is faster.

Alveoli

In the human lung, each tube ends in a cluster of alveoli. This is where oxygen and carbon dioxide gases are exchanged with the blood. There are millions of alveoli in each lung.

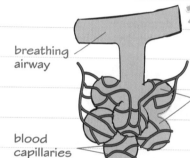

breathing airway

air moves in and out

alveoli greatly increase the surface area for diffusion which makes the exchange of gases more effective

blood capillaries

The diagram shows two villi in the human small intestine. Explain **three** ways the small intestine is adapted to make absorption of substances as efficient as possible.

(6 marks)

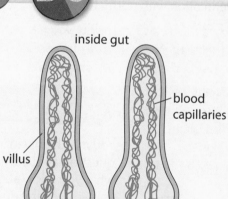

inside gut

blood capillaries

villus

inside body

The large numbers of villi increase the surface area of the small intestine for absorption. The extensive network of capillaries carries absorbed food molecules quickly away to the rest of the body. This maintains a high concentration gradient.

The single layer of cells that cover the villi provides a short diffusion path from the food in the gut to the blood vessels.

Now try this

target **G-E**

1 Name **one** human organ that is specially adapted for exchanging substances with the environment. *(1 mark)*

target **D-C**

2 State **one** way in which alveoli help to increase the effectiveness of exchange of gases. *(1 mark)*

target **B-A**

3 Explain why an extensive capillary network and ventilation of the lungs helps to maximise the effectiveness of gas exchange. *(4 marks)*

Ventilation

The lungs are part of the breathing system. The breathing system takes air into and out of the body. In the lungs:
- oxygen diffuses from the air into the blood
- carbon dioxide diffuses from the blood into the air.

VENTILATION means taking air into and out of the lungs:

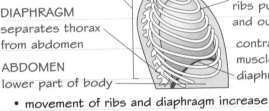

RIBS
protect lungs (and heart)

THORAX
upper part of body

DIAPHRAGM
separates thorax
from abdomen

ABDOMEN
lower part of body

← air in

contraction of
muscles between
ribs pulls ribs up
and out

contraction of
muscle flattens
diaphragm

air out →

relaxation of muscles
between ribs lets ribs
move down and in

relaxation of muscle
allows diaphragm
to dome upwards

- movement of ribs and diaphragm increase thorax volume
- pressure inside lungs decreases
- air pressure outside greater than in lungs
- air drawn in through mouth

- movements of ribs and diaphragm decrease thorax volume
- pressure inside lungs increases
- air pressure in lungs greater than outside
- air forced out of mouth

Worked example

 target B–A*

 AQA SKILL Compare Page 79

If a person is not able to breathe on their own after an accident, a ventilation mask may be placed over their nose and mouth. When the gas in the mask is at high pressure, it forces air into the lungs. As the pressure is reduced, the muscles between the ribs relax and air leaves the lungs. Compare this artificial ventilation with normal breathing. *(3 marks)*

Normally, breathing in is caused by an increase in thorax volume as muscle contraction pulls the ribs up and out and flattens the diaphragm. The decrease in pressure in the thorax draws air in. This is different from using the mask, which pushes air in.

In normal breathing air is forced out of the lungs, caused by an increase in pressure as the muscles relax, moving the ribs and diaphragm so that the thorax volume decreases. This is similar to using the mask.

Anything with this Skills sticker is helping you to apply your knowledge.

 AQA SKILL Explain Page 79

EXAM ALERT!

You may be given information in an exam question about artificial breathing aids. Apply what you know about normal ventilation to help you answer any questions like this.

Students have struggled with exam questions similar to this – **be prepared!**

Now try this

 target G–E

1 Describe the movement of the ribs when you breathe in deeply. *(1 mark)*

 target D–C

2 Explain whether the volume of the thorax is greater during breathing out or breathing in. *(3 marks)*

 target B–A*

3 An iron lung is an old form of artificial ventilator. The patient lay inside a large, airtight metal case, with only their head outside. The pressure inside the ventilator was increased to increase the pressure on the patient's lungs. The pressure was then decreased. Explain how the iron lung caused ventilation of the lungs. *(4 marks)*

Exchange in plants

Leaves

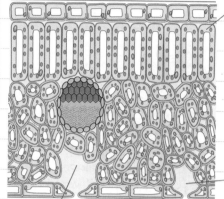

flattened shape of leaf gives large surface area

guard cells change shape to open or close a stoma

open   closed

Water vapour is also lost from leaves through the stomata. This is called TRANSPIRATION.

stomata (pores) allow carbon dioxide from air into leaf and allow oxygen from photosynthesis to leave leaf

internal air spaces increase surface area for diffusion of gases

One **stoma**, two or more **stomata**.

Roots

root hairs greatly increase the surface area for absorbtion

root tip

Most of the water and mineral ions in a plant are absorbed through the roots.

Water balance

If water evaporates faster from the leaves than it is taken up by the roots, the plant will start to WILT (go floppy).

The guard cells can change shape to close the stomata to prevent further wilting.

Worked example

target D-C

AQA SKILL Explain Page 29

Three identical potted plants were watered and weighed. Their pots were wrapped and sealed in plastic bags so that only the plant was in the open air. Each plant was then placed in different conditions. After 6 hours, the bags were removed and the plants weighed again. The table shows the results.

(a) Calculate the percentage change in mass for plants B and C. *(2 marks)*

(b) Explain the differences in results. *(2 marks)*

Remember that transpiration rate also increases in dry conditions.

Mass in g	Plant A (cool still air)	Plant B (warm still air)	Plant C (warm windy)
at start	436	452	448
at end	412	398	332
% change	5.5	11.9	25.9

The results show that evaporation of water from the plant was faster in warm air than cool air and even faster in windy air than in still air. This is because evaporation from the stomata is faster in hot and windy conditions.

Now try this

target D-C

target B-A*

1 Describe **two** ways in which plant leaves are adapted for the exchange of gases with the air. *(4 marks)*

2 A pot plant placed near a warm radiator wilted much more quickly than when it was placed away from the radiator.

(a) Suggest a reason for this difference. *(2 marks)*

(b) A student looked at the underside of the leaf of the wilted plant with a hand lens. Explain whether the student would see open or closed stomata. *(2 marks)*

(c) State how the size of the stomata is controlled. *(1 mark)*

The circulatory system

The circulatory system consists of the heart, blood vessels and blood. The role of the circulatory system is to transport substances around the body.

Structure of the heart

The heart pumps blood around the body.

The heart is always drawn as if you were looking at a person. So the right side of the heart is on the left side of the diagram.

One **atrium**, two **atria**.

The walls of the heart chambers are mostly made of muscle.

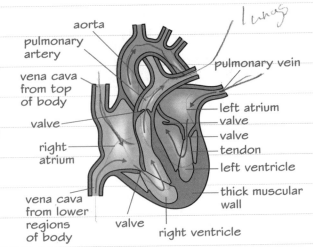

aorta
pulmonary artery
pulmonary vein
vena cava from top of body
valve
right atrium
left atrium
valve
valve
tendon
left ventricle
thick muscular wall
vena cava from lower regions of body
valve
right ventricle
lungs

Blood circulation

Valves in the heart make the blood flow in the right direction.

| Blood enters the atria. | → | The atria contract, forcing blood into the ventricles. | → | The ventricles contract, forcing blood into the arteries. | → | Blood flows through arteries to the organs and returns to the heart through veins. |

There are two circulation systems: one through the lungs and one through all the other organs.

 Worked example   target **D-C**

 AQA SKILL Compare Page 79

Patients with heart failure need an urgent transplant to give them a replacement heart. Patients who have had heart transplants may survive for many years, but getting a transplant depends on a suitable heart being available after the death of another person. If no suitable heart is available, the patient may be given an artificial heart. Artificial hearts have a limited use of a few years.

Compare the usefulness and drawbacks of artificial hearts.   (2 marks)

An artificial heart is useful because it keeps the patient alive until a suitable human heart is available. The disadvantage of getting an artificial heart is that it doesn't last that long and a suitable human heart must be found in the time that the artificial heart

Faulty heart valves may allow blood from the two circulations to mix. This means that blood flowing to the body may not carry as much oxygen. Faulty valves can be replaced with artificial ones.

Now try this

target **G-E**

1 Name the heart chamber that receives blood from the lungs.
(1 mark)

target **D-C**

2 State why there are valves in the heart.   (1 mark)

target **B-A***

3 The muscular wall of the left ventricle of the heart is much thicker than that of the right ventricle. Suggest a reason for this difference.
(3 marks)

Blood vessels

Arteries and veins

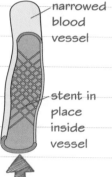

space where blood flows

thick wall of muscle and elastic fibres

artery (cross-section)

large space for blood to flow

thinner wall than artery

vein (cross-section)

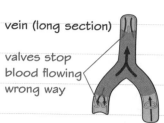

vein (long section)

valves stop blood flowing wrong way

Stents

If an artery becomes narrowed, the drop in blood flow can cause damage to tissue beyond the blockage. A collapsed wire frame, called a STENT, is inserted into the narrowed part of the artery. The stent is expanded using a small balloon. The balloon is then removed.

narrowed blood vessel

stent in place inside vessel

Stents are often used in **coronary** arteries that supply the heart muscle with nutrients and oxygen.

Exchange in capillaries

Substances are exchanged between the body cells and blood in CAPILLARIES.

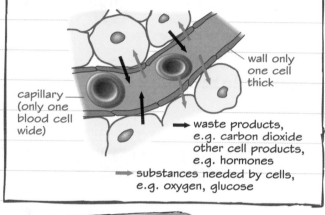

wall only one cell thick

capillary (only one blood cell wide)

→ waste products, e.g. carbon dioxide other cell products, e.g. hormones

→ substances needed by cells, e.g. oxygen, glucose

EXAM ALERT!

Remember to calculate percentages so you can make a fair comparison.

Students have struggled with exam questions similar to this – **be prepared!**

Worked example

  B–A*

AQA SKILL Evaluate Page 79

In a study of patients with narrowed coronary arteries, one group exercised and was given medicine, while a similar group exercised, took medicine *and* had a stent inserted into the narrowed artery.

In the 'without stent' group, 202 out of 1092 patients died of a heart attack within 5 years. In the 'with stent' group, 211 out of 1111 died of a heart attack within 5 years. Do these results suggest that stent surgery is worth doing? *(3 marks)*

The percentage of deaths was 18.5% in the group without stents and 19.0% for the group with stents. This is not a big difference with such large sample sizes. This suggests that stent surgery is not worth doing if other treatments are carried out at the same time.

Now try this

 G–E

1 Choose the correct word from the word box to answer these questions.

| artery   capillary   vein |

(a) Which type of blood vessel is the narrowest? *(1 mark)*

(b) Which type of blood vessel has the thickest wall? *(1 mark)*

 D–C

2 Explain why a stent may be inserted into a narrowed artery. *(2 marks)*

B–A*

3 Explain why almost every body cell is very close to a capillary. *(4 marks)*

Blood

Blood is a tissue, made of several types of cell suspended in a fluid called PLASMA.

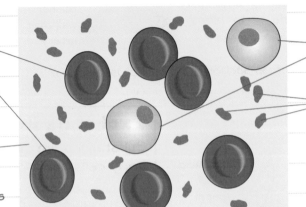

red blood cells
- no nucleus
- packed with red haemoglobin

liquid plasma transports
- carbon dioxide from organs to lungs
- soluble digested food molecules from small intestine to other organs
- urea from liver to kidneys

white blood cells
- have a nucleus
- help body defend against infection by microorganisms

platelets
- small cell fragments with no nucleus
- help blood to clot at a wound

Haemoglobin

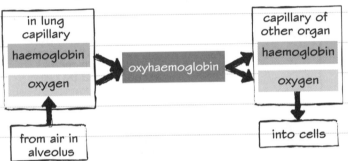

in lung capillary
haemoglobin
oxygen

from air in alveolus

oxyhaemoglobin

capillary of other organ
haemoglobin
oxygen

into cells

Worked example

AQA SKILL Evaluate Page 19

target **B-A***

A person who has lost a lot of blood needs a blood transfusion urgently to prevent death. Human blood can only be given to the person if it is of the right type, is free from infection and has been stored properly. Artificial blood can be made that carries out many of the functions of human blood. It can be stored for longer and is sterile, but it can sometimes cause problems, including death. Evaluate the use of artificial blood for transfusions.   *(2 marks)*

Artificial blood is useful when there is no suitable human blood available as it will stop the person dying. But if there is suitable human blood, it should be used instead to avoid problems that may be caused by the artificial blood.

Now try this

target **G-E**

1 What is the function of **white blood cells**?
          (1 mark)

target **D-C**

2 Artificial blood contains chemicals that can bind with and release oxygen. Explain how this is similar to normal human blood.   *(3 marks)*

target **B-A***

3 The US army has paid for research into the development of artificial blood. Suggest **three** reasons why artificial blood could be very useful in combat areas.
          (3 marks)

Transport in plants

Plants have two separate transport systems: XYLEM and PHLOEM.

> Xylem and phloem are grouped together in plant veins.

Xylem

The movement of water from the roots to the leaves of a plant through the xylem is called the TRANSPIRATION STREAM.

> Soluble mineral ions are also absorbed from the soil and transported around the plant in xylem tissue.

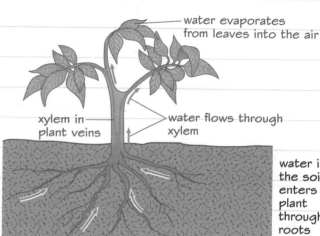

water evaporates from leaves into the air

xylem in plant veins

water flows through xylem

water in the soil enters plant through roots

Phloem

Dissolved sugars are transported around a plant in phloem.

dissolved sugars needed for growth in growing regions, e.g. bud

sugars are produced in the leaves after photosynthesis

phloem in plant veins

dissolved sugars are carried around the plant in phloem

storage organ (e.g. potato)

dissolved sugars may be stored in storage organs so they can be used later

---

Worked example   D-C

One end of a stalk of celery was placed in a beaker of water containing red dye for an hour. The stalk was then cut straight across every 2 cm from its base. The presence or absence of red dye in the veins at each cut was recorded. The table shows the results.

Distance in cm	0	2	4	6	8	10
Dye visible?	yes	yes	yes	yes	yes	no

(a) Calculate the rate of movement of the dye. *(2 marks)*

Distance moved = 8 cm
Time = 1 hour
Rate of movement = 8 cm/h

(b) Explain why it moved. *(2 marks)*

The dye moved with the water in the transpiration stream through the xylem in the stalk.

Now try this

 G-E D-C

1 Name the **two** transport systems found in plants. *(2 marks)*

2 An aphid is an insect pest that inserts needle-like mouthparts into the veins of a plant. It feeds on sugars.
Explain which tissue in the vein the aphid is feeding from. *(2 marks)*

 B-A*

3 A young potato plant grows rapidly during the spring and summer, but in the autumn develops new potatoes, which are storage organs. Compare the direction of movement of sugars in the plant at different times of the year. *(3 marks)*

Biology six mark question 1

There will be one six mark question on your exam paper, which will be marked for *quality of written communication* as well as scientific knowledge. This means that you need to apply your scientific knowledge, present your answer in a logical and organised way and make sure that your spelling, grammar and punctuation are as good as you can make them.

Worked example

Plant cells need nitrogen ions to make proteins for growth. Plants cannot take nitrogen from the air, but they can absorb soluble nitrogen ions from the soil.

Describe as fully as you can the cells, tissues and processes involved in getting nitrogen ions in the soil to growing plant cells. *(6 marks)*

Plants absorb nitrogen ions from the soil though root hair cells. These cells are specially adapted with long extensions to increase their surface area for absorption.

In the plant the ions move into the xylem tissue, which carries them up the plant and to the growing areas.

EXAM ALERT!

Always plan what you are going to write for the six mark questions. You are given credit for a well-organised answer.

Students have struggled with exam questions similar to this – **be prepared!**

This part of the answer has failed to mention that the nitrogen ions are taken up by active transport. This is because the concentration of nitrogen ions is lower in the soil than in plant cells. So energy is needed to move the ions up their concentration gradient into the plant.

This part of the answer would have been better if it also said that the ions are carried with water as part of the transpiration stream between the roots of the plant and the leaves.

Apply your knowledge

Be confident when answering questions that give examples you have not studied previously. Use the information in the question and apply the knowledge and understanding of the topic that you already have.

Now try this

Respiring cells need oxygen. In humans, oxygen is exchanged between the body and the air. Describe the cells, tissues and processes involved in getting oxygen from the air to respiring cells.

(6 marks)

Removing waste products

Waste products from cell processes must be removed from the body before they cause harm. Carbon dioxide and urea are two waste products made in the body.

carbon dioxide → produced in cells by respiration → removed from body through the lungs when we breathe out

urea → produced in the liver from breakdown of amino acids → removed from body by kidneys in urine ← Urea is toxic and will damage the body if it builds up in large quantities.

Worked example D-C

(a) State how water and ions enter the body.   *(1 mark)*

Water and ions enter the body in our food and drink.

(b) State why the water and ion content of the body is controlled.   *(1 mark)*

The water and ion content of the body must be controlled because otherwise too much water may move into or out of cells, and that would damage them.

Processes in the kidney

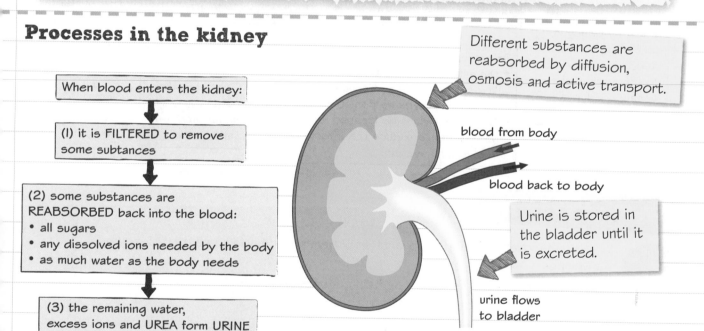

When blood enters the kidney:

↓

(1) it is FILTERED to remove some subtances

↓

(2) some substances are REABSORBED back into the blood:
- all sugars
- any dissolved ions needed by the body
- as much water as the body needs

↓

(3) the remaining water, excess ions and UREA form URINE

Different substances are reabsorbed by diffusion, osmosis and active transport.

blood from body

blood back to body

Urine is stored in the bladder until it is excreted.

urine flows to bladder

Now try this

 G-E

1 Choose the correct word from the word box to answer these questions.

| heart | kidney | liver | stomach |

(a) In which organ is urea formed?   *(1 mark)*

(b) In which organ is urea removed from the blood?   *(1 mark)*

D-C

2 Explain what the **two** main processes carried out in the kidney are.   *(4 marks)*

 B-A*

3 Explain why the kidneys use active transport to reabsorb sugar and ions but water passes back into the blood through osmosis.   *(4 marks)*

Kidney treatments

People with kidney failure may be treated with DIALYSIS, or by TRANSPLANT of a healthy kidney.

Dialysis

Dialysis must be carried out every 2 or 3 days, usually in a hospital.

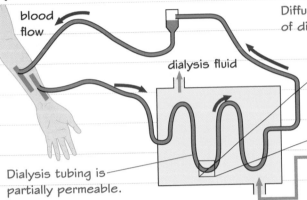

blood flow

dialysis fluid

Diffusion restores the normal concentrations of dissolved substances in the blood.

Urea diffuses out of the blood into the fluid.

Dialysis tubing is partially permeable.

Dialysis fluid contains the same concentration of useful substances as blood so glucose and useful mineral ions are not lost.

Kidney transplant

A healthy kidney is connected to the blood circulation, to do the work of the diseased kidneys. :)

Problem: The **antigens** on the transplanted kidney cells are different from antigens on cells in the patient's body.

The **antibodies** in the patient's immune system attack the transplanted kidney and **reject it**. :(

To prevent rejection:
• the antigens on the transplanted kidney and patient's tissues must be as similar in type as possible
• the patient must be treated for life with drugs to reduce the effects of the immune system.

This means the patient may get more infections than normal. :(

Antigens are proteins on the surface of cells. They are the same on all cells from one person, but differ from person to person.

target B-A*

AQA SKILL Evaluate Page 79

Evaluate the advantages and disadvantages of treating kidney failure by dialysis and by transplant. *(4 marks)*

Both treatments help to keep the balance of substances in the blood. If this didn't happen the person would soon die.

Dialysis is a problem because it has to be done every few days, usually in a hospital, which takes a lot of time. A transplant is a problem because the person has to take drugs to suppress their immune system, which means they may get more infections.

Transplant is better for most patients than dialysis because they don't have to go back to hospital every few days.

Now try this

target G-E

target D-C

1 Name **two** ways that kidney failure is treated. *(2 marks)*

2 Describe how dialysis works to make sure that the blood has the right concentration of substances. *(4 marks)*

target B-A*

3 Describe the role of the immune system in transplant rejection. *(3 marks)*

Body temperature

The temperature in the core (centre) of the body needs to be kept within a narrow temperature range so that cells can work properly.

Sweating

One way we keep cool is by SWEATING.

| As the body gets hotter, more sweat is produced to cool the body. | → | More sweat means more water is lost. | → | More water must be taken in as food or drink to replace the lost water. |

> When the body gets hot, pale skin becomes pinker as more blood flows through the skin.

The thermoregulatory centre

The THERMOREGULATORY CENTRE in the brain monitors and controls core body temperature.

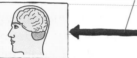

thermoregulatory centre in the brain

impulses sent to thermoregulatory centre

RECEPTORS in thermoregulatory centre detect temperature of blood in brain

receptors in skin detect temperature of skin

> Core body temperature is the temperature in the middle of the body, e.g. around the heart and liver. Parts of the body further from the core may be much cooler.

HIGHER

Cooling down

If core body temperature is too high:

- blood vessels supplying the skin capillaries DILATE (get wider)
- more blood flows through the skin capillaries near the skin surface, so more heat is transferred to the cooler environment
- sweat glands release more sweat – the sweat evaporates transferring heat to the environment.

Worked example D-C HIGHER

Explain **two** changes that occur in the body if core temperature is too low.   *(2 marks)*

1 Blood vessels that supply skin capillaries constrict and get narrower, which reduces blood flow through the capillaries near the skin surface. This means that less heat is transferred to the environment.

2 Muscles may start to shiver, which releases energy to warm the body.

Now try this

1 Name the part of the body that monitors and controls body temperature.   *(1 mark)*

2 Explain why we need to drink more on a hot, sunny day.   *(2 marks)*

3 Explain how **two** responses of the body to being too cold help to restore normal core body temperature.   *(4 marks)*

HIGHER

Blood glucose control

The concentration of glucose in the blood is monitored and controlled by the PANCREAS.

Insulin

INSULIN is a hormone.

> blood glucose concentration is too high, e.g. just after a meal

⬇

> insulin is released by the pancreas into the blood

⬇

> insulin causes cells (especially in muscles and the liver) to take up glucose from the blood

Glucagon HIGHER

GLUCAGON is another hormone.

> blood glucose concentration is too low, e.g. hours after a meal

⬇

> glucagon is released by the pancreas into the blood

⬇

> glucagon causes glycogen stores to be broken down to glucose, which is released into the blood

> Glucose is stored in cells as glycogen.

Type 1 diabetes

Type 1 diabetes is caused by the pancreas not producing enough insulin.
This means the blood glucose concentration can rise to a dangerously high level.
It can be treated with injections of insulin balanced with careful attention to diet and exercise.

> It is dangerous for blood glucose concentration to rise too high or fall too low.

Worked example

B-A*

AQA SKILL Explain Page 79

People who have type 1 diabetes inject insulin to control blood glucose concentration. Injected human insulin is not absorbed well, so new forms of insulin are being developed. The graph shows the blood glucose concentration of people with Type 1 diabetes using a new form of insulin compared with those using human insulin. Explain whether the graph shows that the new form of insulin is better.

(2 marks)

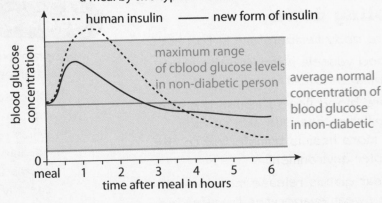

Control of blood glucose after a meal by two types of insulin

----- human insulin   —— new form of insulin

blood glucose concentration

maximum range of cblood glucose levels in non-diabetic person

average normal concentration of blood glucose in non-diabetic

0 | meal | 1 | 2 | 3 | 4 | 5 | 6
time after meal in hours

The new form of insulin is better than human insulin because it stops blood glucose concentration increasing as high or falling as low as using human insulin.

Now try this

target **D-C** HIGHER

1 State the roles of insulin and glucagon in controlling blood glucose concentration.   *(2 marks)*

target **B-A***

2 Carbohydrates in food are digested to glucose. Suggest why the dose of insulin injected by a diabetic may vary depending on what they eat.   *(3 marks)*

Biology six mark question 2

There will be one six mark question on your exam paper, which will be marked for *quality of written communication* as well as scientific knowledge. This means that you need to apply your scientific knowledge, present your answer in a logical and organised way and make sure that your spelling, grammar and punctuation are as good as you can make them.

Worked example

People with kidney failure are first treated with dialysis. Discuss the different ways in which the kidneys and dialysis keep the right concentrations of substances in the blood, and why dialysis must be done on a regular basis.

(6 marks)

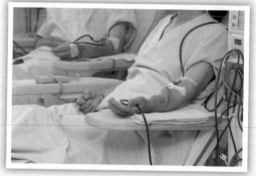

The kidneys first filter things out of the blood then the blood reabsorbs what it needs for the body to stay healthy. What is left over forms urine, which is excreted.

In dialysis, urea and other substances diffuse out of the blood through partially permeable membranes into the dialysis fluid.

Urine contains urea, which is a waste product made in the liver from the breakdown of amino acids. Urea is toxic in the body, so must be removed on a regular basis to stop it building up.

Discuss

This question expects you to write down the similarities and differences between the way that the kidneys work and the way that dialysis works. Make sure you write about both methods, not just one.

This part of the answer would have been much better if it had stated that water, glucose, urea and mineral ions are filtered out of the blood, and that all glucose and any water and mineral ions that are needed are reabsorbed.

This part of the answer should have explained that the concentrations of substances in the dialysis fluid are controlled so that the normal levels of substances in the blood are restored at the end of treatment.

Now try this

Patients with Type 1 diabetes need to inject insulin regularly to control blood glucose concentration. Explain how blood glucose concentration is normally controlled by insulin, and why factors such as diet and exercise must be considered when judging the dose of insulin to inject.

(6 marks)

Pollution

POLLUTION causes harm to the environment and the organisms living in it.

Increasing pollution

| rapid growth in human population | + | increased standard of living | = | increasing waste produced | → | if not handled properly, causes more pollution |

↓

uses more land for:
- building
- quarrying for building materials
- farming to produce food
- dumping waste

→ less space for other animals and plants

Our **standard of living** is affected by all the products we use and make.

Examples of pollution

Pesticides are used on crops to kill pests. **Herbicides** are used on crops to kill weeds.

POLLUTION →

of the air:
- smoke and gases, e.g. sulfur dioxide, which is part of acid rain

Sewage contains waste water from cleaning etc. as well as human waste.

of the land:
- toxic chemicals, e.g. pesticides and herbicides

of the water:
- sewage
- fertiliser
- toxic chemicals washed in from land

Worked example

D-C

AQA SKILL
Explain
Page 79

In the UK, power stations are given limits on how much sulfur dioxide they can release into the air. The amount they are allowed to release has been reduced over the past 20 years.
The graph shows the amount of sulfur dioxide in the air. Explain whether the graph suggests that the limits are working. *(2 marks)*

The amount of sulfur dioxide in the air is decreasing, so the limits seem to be working.

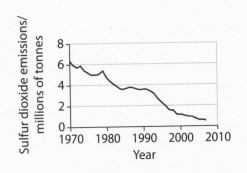

Now try this

G-E

D-C

1 Give **one** example of water pollution.
(1 mark)

2 A new quarry is planned for an area of wild moorland. Describe **two** negative effects the new quarry might have on the area. *(4 marks)*

B-A

3 Explain why pollution by some pesticides and herbicides is increasing. *(2 marks)*

Deforestation

Large-scale DEFORESTATION (cutting down of trees) is taking place in tropical areas.

Deforestation provides ...

- land cleared for agriculture

- timber, e.g. for building

- to grow food crops, e.g. rice

- grassland for feeding cattle

- to grow crops for making biofuels (see page 19)

Methane is released from cattle and rice fields into the air.

EXAM ALERT!

Be prepared to make the connections between deforestation and an increase in methane and carbon dioxide in the air.

Students have struggled with this topic in recent exams – **be prepared!**

Deforestation leads to ...

- burning of trees

- increased activity of microorganisms that decay chopped wood

→ more carbon dioxide released into air

- fewer trees so less carbon 'locked up' for many years in wood

→ less carbon dioxide removed from air

Peat

Peat, e.g. from peat bogs, is used:

- as a fuel in some places
- to make garden composts.

Destruction of peat releases carbon dioxide into the air.

Peat-free composts are being made, so that the destruction of peat is reduced.

Worked example   B-A*

AQA SKILL — Describe — Page 19

A study of ground-living arthropods (e.g. beetles, millipedes) in Borneo found between 7 and 11 different arthropod groups in original rainforest, and 4 or 5 arthropod groups in oil palm plantations that have replaced the rainforest. Use this data to describe the effect of replacing rainforest with plantation trees of one type on the biodiversity there. *(3 marks)*

The study shows that biodiversity in terms of the number of groups of arthropods is smaller in the plantations than in the original rainforest. The data do not show if the same groups are found in each place or if they are different. Reducing the number of arthropod groups will affect other organisms, such as animals that prey on them for food.

Now try this

G-E

1 Give **one** reason why there is deforestation of tropical trees. *(1 mark)*

D-C

2 Explain why gardeners should be encouraged to buy peat-free composts instead of peat-based composts for their gardens. *(2 marks)*

 B-A*

3 Explain why deforestation can lead to an increase in carbon dioxide in the air. *(3 marks)*

Global warming

Increasing concentrations of carbon dioxide and methane in the air are contributing to GLOBAL WARMING (an increase in Earth's temperature).

An increase in Earth's temperature of just a few degrees Celsius could …

- cause big changes in climate

  e.g. drought, colder winters, hotter summers, flooding

- cause a rise in sea level

  due to ice melting and to warmer sea water (expands as it gets warmer)

- affect species

  e.g. change in migration of birds, change in distribution of species (the area they live in)

Carbon dioxide in water

Carbon dioxide is removed from the air by trees, but can also be SEQUESTERED in bodies of water.

To sequester means to isolate or hide away. In this case the carbon dioxide is 'hidden away' from the normal cycling of carbon between organisms and the air.

large amounts of carbon dioxide from the air dissolve in water, such as in ponds, lakes and oceans

some is converted to:
- chemicals in sediments

- chemicals in organisms, such as the shells of shellfish

Worked example

target B-A*

AQA SKILL
Evaluate
Page 79

Bluebells flower in spring. The table shows the date of the first bluebell flower recorded at a particular site in certain years.

	1996	2000	2004	2008	2012
first flower of bluebell	23 April	13 April	14 April	10 April	4 April

(a) Evaluate the method used to collect the data. *(1 mark)*

The data were collected at the same site, so that increases the reliability of the results because other factors that change from site to site will be controlled.

(b) Evaluate the data as evidence for climate change. *(2 marks)*

The dates given in the table are getting earlier, but the table only shows some years. You would need a date for every year to be sure whether or not there was a trend.

Now try this

1 Name **one** way in which global warming could affect climate. *(1 mark)*

2 Climate change may make it possible for tropical diseases to spread to other areas. Suggest how this might happen.

(2 marks)

3 Carbon dioxide dissolves less well in warmer water, and is not as easily absorbed by organisms such as shellfish. Explain the importance of this for global warming. *(4 marks)*

Biofuels

BIOFUELS are fuels made from the fermentation of plant or animal materials.

Biogas

Mostly methane. ➡ BIOGAS is a gas produced by ANAEROBIC fermentation of materials that contain carbohydrates. ⬅

Suitable carbohydrates are found in animal waste and in a wide range of plant materials, including plant waste left after crops have been harvested.

Fermentation caused by microorganisms that do not use air for respiration.

Generators

Biogas can be generated on a large commercial (industrial) or small domestic (family) scale.

Commercial scale generation	Domestic scale generation
• wide range of waste materials from agriculture used, e.g. crop stalks • biogas used to generate electricity • leftover solids and liquids are sold as fertiliser	 biogas burned in kitchen biogas manure anaerobic fermentation produces biogas biogas generator dried leftover waste used as fertiliser

 Worked example **D-C**

AQA SKILL
Explain
Page 19

In colder countries, commercial biogas generators are insulated and heated to keep them warm, while in hot countries biogas generators are often built underground. Explain this difference.
(2 marks)

Biogas is produced by the fermentation of carbohydrates by microorganisms. Microorganisms can grow and digest materials faster if they are kept warm. However, they may be killed if it gets too hot, so building the generator underground in hot places can help stop this happening.

Now try this

G-E

1 Choose the correct words from the box to complete the sentences.

| aerobic   anaerobic   carbon dioxide   methane   oxygen |

Biogas is made by the _____ fermentation of carbohydrates. Biogas mostly contains the gas _____.
(2 marks)

D-C

2 State **two** differences between a commercial and a domestic biogas generator. *(2 marks)*

B-A*

3 Commercial biogas generators are being built to replace some power stations that run on fossil fuels. Suggest how biogas could help to reduce the problem of increasing carbon dioxide concentration in the air.
(4 marks)

Food production

FOOD PRODUCTION is the growing and preparing of food for sale in the shops.

More energy is available for humans if we eat the plants, rather than eating the animals that eat the plants. For example ...

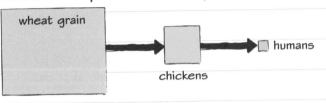

wheat grain → chickens → humans

> Remember that energy and biomass are lost to the environment at each stage of a food chain, as heat energy and carbon dioxide from respiration, and as chemical energy in dead biomass (and in waste from animals).

wheat grain → humans

The boxes represent energy or material in the biomass of the organisms.

Factory farming

In factory farming, food animals are kept in conditions that help them grow as fast as possible.

| food animals lose energy:
• by moving around
• as heat to the environment | → | less energy is lost if:
• movement is restricted, e.g. animals are kept in cages
• the environment is kept warm | → | ✓• the animals grow faster and bigger
✗• animals close together are more likely to get ill
✗• heating uses electricity
✗• it is unethical to keep animals in stressful or unhealthy conditions |

> In a question on food production, you may need to consider the pros and cons of factory farming using ideas like these.

Worked example   target D-C

AQA SKILL
Explain
Page 79

The 'Freedom Food' logo can only be used on foods produced to particular standards. For example, a maximum of 19 chickens can be kept per square metre of barn. In factory farming, the density of chickens is higher than this. Freedom Food chicken is more expensive than factory-farmed chicken.

Give **one** advantage and **one** disadvantage to a farmer of keeping chickens to Freedom Food standards.   (2 marks)

The advantage is that the farmer will get more money for each chicken that is sold. The disadvantage is that the farmer won't be able to keep as many chickens in the same space.

Now try this

target D-C

1 Explain how factory farming can increase the rate at which food animals can grow.   (3 marks)

target B-A*

2 Look at the following food chain from the sea.

microscopic plankton → sardine (small fish) → tuna (large fish)

In terms of efficiency of food production, explain at which level of this food chain humans should gather food from the sea.   (4 marks)

Fishing

We catch fish for food. Many wild fish stocks are decreasing because we are OVERFISHING (taking too many fish).

Many sizes of fish are caught by fishermen. If there aren't enough large fish in the catch, they may also take smaller fish. The rest of the catch is usually thrown back into the water.

Fish take several years to grow large enough to breed.

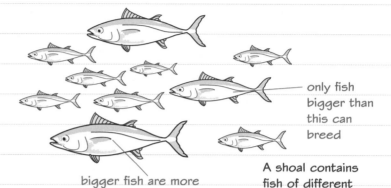

only fish bigger than this can breed

bigger fish are more expensive than smaller fish

A shoal contains fish of different sizes and ages.

AQA SKILL
Explain
Page 19

| If fishing takes all the large fish ... | → | there are no fish left that are big enough to breed ... | → | so no young fish replace the fish taken in fishing ... | → | the fishery collapses (there are not enough fish to make fishing worthwhile). |

- In the short term, there are plenty of fish for us to eat.
- In the long term, there will be no fish for us to eat in the future. This level of fishing is not SUSTAINABLE.
- CONSERVATION (protection) of fish stocks is needed so we have fish to eat in the future.

Fishing quotas

Fish stocks can be conserved by giving fishermen limits on catch size, called QUOTAS. They are not allowed to catch more than the quota. This stops too many fish being taken from an area.

Worked example   B-A*

Explain how limits on net size can help to conserve fish stocks. *(3 marks)*

The mesh size of nets is limited by a minimum size. This makes sure that small fish are not caught. If the mesh limit is larger than the smallest size for a breeding fish, this will make sure there are always some breeding fish left in the area to produce new fish.

EXAM ALERT!

You should know that in order for fishing to be sustainable enough adult fish have to be left to breed each year.

Students have struggled with exam questions similar to this – **be prepared!**

Now try this

 G-E

1 Give **one** reason why we may not have any fish of some species to eat in the future. *(1 mark)*

 D-C

2 Explain **two** ways in which fish stocks can be conserved. *(2 marks)*

target B-A*

3 Every year the European Union sets fishing quotas for all its member countries. The quotas are based on scientific advice about fish stocks. However, the quotas set by the politicians are often at higher levels than the scientists advise. Explain a reason for this. *(2 marks)*

Sustainable food

MYCOPROTEIN is a protein made from a fungus.

Making mycoprotein

| *Fusarium* fungus is grown on glucose syrup in AEROBIC conditions. | → | The fungus biomass is harvested and purified. | → | The biomass is used to make protein-rich foods. |

These foods are suitable for vegetarians to eat because they are not made by killing animals.

'Aerobic' means 'using air'.

Mycoprotein is made from large amounts of a microscopic fungus, *Fusarium*. It can be processed to resemble chicken or beef and is a relatively new method of food production.

Food distribution

Food needs to be transported from where it is grown and prepared for sale to the shops where people buy it.

The distance food travels to the shop is known as FOOD MILES.

Food miles are a concern because transport uses fossil fuels to power the engines. Burning fossil fuels contributes to air pollution and adds carbon dioxide to the air. Some people prefer to buy foods that have been produced as close as possible to the shops.

Worked example

 AQA SKILL Explain Page 79

Plants need enough water to grow well. In areas where there is not enough rainfall, crops may be irrigated. A study looked at three ways of irrigating rice plants – constant irrigation, watering only when the ground was a little dry, and watering where the ground was very dry.

Explain which method of irrigating is the best. *(2 marks)*

Values compared with constant watering	Watering rice crop when a little dry	Watering rice crop only when very dry
Amount of water used	−56%	−75%
Grain yield	+9.4%	−7.6%

Watering when a little dry produces the greatest yield of crop. It also uses around half the water that is needed for constant irrigation, so it wastes less water.

You need to understand that water is a valuable resource and needs to be carefully managed.

Now try this

1 What is meant by **food miles**? *(1 mark)*

2 Explain why mycoprotein is suitable for vegetarians. *(2 marks)*

3 Suggest **four** conditions and why they are needed in the vats used for growing the fungus used to make mycoprotein. *(4 marks)*

Biology six mark question 3

There will be one six mark question on your exam paper, which will be marked for *quality of written communication* as well as scientific knowledge. This means that you need to apply your scientific knowledge, present your answer in a logical and organised way and make sure that your spelling, grammar and punctuation are as good as you can make them.

Worked example

Protein is an essential part of the human diet. The table shows some costs of producing different protein-containing foods.

Food (all values are per tonne of food produced)	Land area in ha	Contribution to global warming in kg CO$_2$ equivalent	% protein in food
wheat	0.14	804	12.7
beef cows	2.30	15 800	22.5
mycoprotein	0.17	2300	16.3

Use the information in the table to discuss the best way of providing protein for people as the global human population increases. *(6 marks)*

The table shows that you need much more land to grow one tonne of beef than one tonne of wheat or mycoprotein. Producing beef also contributes more to global warming than producing mycoprotein, though producing wheat contributes the least. But there is most protein in beef and least in wheat.

As more people need more protein to eat, the best way to produce that protein would be to grow wheat or mycoprotein as they use less land and do less to cause global warming. Mycoprotein might be a bit better than wheat because it contains more protein.

Reach a conclusion

This question is asking for a conclusion, so you will need to compare the different sets of data and use your comparison to decide the best way to produce protein-rich foods in the long term.

The analysis of the data in the table would be much better if it made numerical comparisons. For example, producing one tonne of beef cow needs 2.3/0.17 = 13.5 times as much land as producing one tonne of mycoprotein.

A better analysis of which food to grow would compare the amount of protein produced with the land area: protein/area is 1.8 for wheat, 9.8 for cows, 2.8 for mycoprotein. This would make it easier to decide that mycoprotein is the best choice.

Now try this

In 2005 around 25% of the world's deforestation took place in the tropical forests of Indonesia. The forests were replaced mainly with oil palm plantations to produce oil for food and for use as a biofuel. Scientists estimated that deforestation contributed about a quarter of global carbon emissions compared with fossil fuels. Describe the benefits and problems with replacing tropical forest with oil palm plantations on this scale.

(6 marks)

The early periodic table

There have been many attempts to classify the elements. John Newlands and Dmitri Mendeleev were two of the scientists who tried to do this in the 19th century.

Information available in the 19th century

In the 19th century, scientists knew:

✓ the properties of the elements

✓ the ATOMIC WEIGHT of the elements.

They did not know about:

✗ protons, neutrons and electrons

✗ atomic numbers (proton numbers).

John Newlands

John Newlands published a table in 1865 containing over 50 elements. His table:

- listed elements in order of atomic weight
- had 7 columns (see right).

Newlands's table was not successful overall:

✓ Counting an element as 1, then counting along, every eighth element had similar properties (Newlands's Law of Octaves).

> Atomic weight is similar to relative atomic mass, A_r.

1	2	3	4	5	6	7
H	Li	Be	B	C	N	O
F	Na	Mg	Al	Si	P	S
Cl	K	Ca	Cr	Ti	Mn	Fe

✗ The Law failed after calcium.

✗ Some metals and non-metals were in the same column, such as oxygen and iron.

✗ There was no room for new elements.

Worked example   D-C

AQA SKILL
Explain
Page 79

Dmitri Mendeleev published a table in 1869. The elements are shown in order of atomic weight. Part of his table is shown on the right.

Mendeleev left gaps in his table (shown as *). Explain why he did this.   *(2 marks)*

		Group					
1	2	3	4	5	6	7	
H							
Li	Be	B	C	N	O	F	
Na	Mg	Al	Si	P	S	Cl	
K	Ca	*	Ti	V	Cr	Mn	
Cu	Zn	*	*	As	Se	Br	
Rb	Sr	Y	Zr	Nb	Mo	*	
Ag	Cd	In	Sn	Sb	Te	I	

Mendeleev's decision to do this was important. The spaces were for elements that had not been discovered, and it let him place elements with similar properties in the same groups.

The modern periodic table contains over 100 elements but far fewer were known then. It is called the periodic table because elements with similar properties are found at regular intervals in it.

Iodine has a lower atomic mass than tellurium so should come first. Mendeleev swapped them round to match their properties better.

Now try this

D-C

1 (a) Which group is missing from Mendeleev's table?   *(1 mark)*

(b) Give **two** similarities, and **one** difference, between Mendeleev's Group 1 and Group 1 in the modern periodic table.   *(3 marks)*

B-A*

2 Suggest **two** reasons why some scientists thought Mendeleev's table was not correct.   *(2 marks)*

The modern table

Atomic structure

The structure of atoms was discovered early in the 20th century. Scientists discovered:

- protons, neutrons and electrons
- that the number of protons in the nucleus gives an element's ATOMIC NUMBER (proton number).

In Mendeleev's time, a periodic table was often seen as a curiosity. An element's atomic number was just its position in the table. Later it was discovered that an atomic number is also the number of protons in a nucleus. The periodic table gradually became a useful tool for describing and predicting properties.

Elements in the periodic table

The elements are arranged in order of atomic number (proton number).

Metals are on the left and non-metals on the right.

Group								
	1	2	3	4	5	6	7	0
Period 1								2 He 2
Period 2	2.1 Li 3	2.2 Be 4	2.3 B 5	2.4 C 6	2.5 N 7	2.6 O 8	2.7 F 9	2.8 Ne 10
Period 3	2.8.1 Na 11	2.8.2 Mg 12	2.8.3 Al 13	2.8.4 Si 14	2.8.5 P 15	2.8.6 S 16	2.8.7 Cl 17	2.8.8 Ar 18
Period 4	2.8.8.1 K 19	2.8.8.2 Ca 20						

After their discovery, Mendeleev added the noble gases as Group 0 in 1902.

The elements in a group have the same number of electrons in their highest occupied energy level (outer shell), except for helium in Group 0.

Worked example

 target D-C

A new element was discovered in 2000. Livermorium has the symbol Lv. Its electronic structure is 2.8.18.32.32.18.6 and its atomic number is 116.

(a) Explain how the atomic number of Lv helps scientists place it in the periodic table.
(2 marks)

The periodic table is arranged in order of atomic number, so Lv should go between elements 115 and 117.

(b) Explain in which group of the periodic table Lv should be placed. *(2 marks)*

Lv should go into Group 6 because its atoms have six electrons in their outer shell.

The **electronic structure** of Lv looks complicated but the last number shows the group into which it should go. The periodic table today is an important summary of the structure of atoms.

Now try this

 target D-C / target B-A*

1 Describe the links between the position of an element in the periodic table and its electronic structure.
(2 marks)

2 Tellurium and iodine have the symbols $^{128}_{52}Te$ and $^{127}_{53}I$. Mendeleev put most of the elements in order of atomic weight but he put tellurium before iodine (even though its atomic weight is greater than iodine's atomic weight). Suggest a reason why he did this, and why it is not a problem in the modern periodic table.
(3 marks)

Group 1

The elements in Group 1 are called the ALKALI METALS.

Properties of Group 1

The alkali metals have these properties. They:

- are metals
- have a low DENSITY (the first three are less dense than water so they float)
- react with non-metals to form ionic compounds that dissolve in water to form colourless solutions
- form ions with a charge of +1
- react with water to form hydrogen and hydroxides. The hydroxides are white solids that dissolve in water to form alkaline solutions.

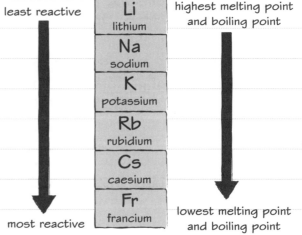

least reactive   highest melting point and boiling point

Li	lithium
Na	sodium
K	potassium
Rb	rubidium
Cs	caesium
Fr	francium

most reactive   lowest melting point and boiling point

Reactions with water

In general:

metal + water → metal hydroxide + hydrogen

For example, for sodium:

sodium + water → sodium hydroxide + hydrogen

$2Na(s) + 2H_2O(l) \rightarrow 2NaOH(aq) + H_2(g)$

> The change in melting and boiling points is not the cause of the change in reactivity.

Worked example   D-C

Sodium and chlorine react together quickly.

(a) Write a word equation for the reaction. *(1 mark)*

sodium + chlorine → sodium chloride

(b) Describe what will happen when the product is added to water. *(1 mark)*

It will dissolve to form a colourless solution.

EXAM ALERT!

You should be able to write word equations for reactions and work out the formulae for ionic compounds. Higher Tier students should also be able to balance symbol equations.

> Students have struggled with exam questions similar to this – **be prepared!**

> The reaction between chlorine and a Group 1 metal becomes more violent as you go down Group 1.

Now try this

1 State **three** properties that the Group 1 elements have in common. *(3 marks)*   (target G-E)

2 Room temperature is around 20 °C. Sodium melts at 98 °C.   (target D-C)

   (a) Suggest a temperature for the melting point of potassium. *(1 mark)*

   (b) Explain why you suggested this temperature. *(2 marks)*

3 (a) Write a word equation for the reaction between lithium and water. *(1 mark)*   (target D-C)

   (b) Hydroxide ions have the formula OH^-. Write the formula for lithium hydroxide. *(1 mark)*

4 (a) State why the Group 1 elements are called the alkali metals. *(1 mark)*   (target B-A*)

   (b) Balance this symbol equation, and include the appropriate state symbols:

   $Rb(s) + H_2O \rightarrow RbOH + H_2$ *(2 marks)*

Transition metals

The TRANSITION METALS have different properties than the alkali metals in Group 1.

Properties of transition metals

Compared to the metals in Group 1 (such as sodium), the transition metals:

- have higher DENSITIES
- have higher melting points and boiling points (except for mercury, which is a liquid at room temperature)
- are stronger and harder
- are much less reactive with water and oxygen.

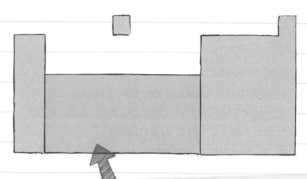

> The transition metals are placed between Groups 2 and 3 in the periodic table.

More properties

Many transition metals also:

- form coloured compounds
- are useful as CATALYSTS
- form more than one type of ion.

> The alkali metals form colourless or white compounds.

> Catalysts alter the rate of a reaction without being used up during the reaction.

> Iron, for example, can form iron(II) ions, Fe^{2+} or iron(III) ions, Fe^{3+}.

Worked example

 D-C

A red-brown oxide of iron has the formula Fe_2O_3.

(a) Explain how you can tell from this information that iron is a transition metal. *(2 marks)*

Transition metals form coloured compounds and iron oxide is described as red-brown.

(b) Give the formula of the iron ion in the oxide. *(1 mark)*

Fe^{3+}

> Copper, for example, is another transition metal. It also forms coloured compounds such as copper(II) sulfate, which is blue in solution.

> Oxide ions have the formula O^{2-}. There are three oxide ions in Fe_2O_3 with a total charge of $(-2 \times 3) = -6$. The charge on the iron ion must be $(+6 \div 2) = +3$.

Now try this

 G-E

1 State **three** properties that the transition metals have in common. *(3 marks)*

2 A student has a white powder and a red powder. Explain how she can decide which one is copper(I) oxide and which one is sodium oxide. *(3 marks)*

 D-C

3 Copper and sodium are good conductors of electricity. Suggest **three** reasons why copper, rather than sodium, is used for electrical wiring. *(3 marks)*

4 Explain why iron is used in the manufacture of ammonia from nitrogen and hydrogen. *(2 marks)*

Group 7

The elements in Group 7 are called the HALOGENS.

Properties of Group 7

The halogens:

- are non-metals
- react with metals to form ionic compounds that dissolve in water to form colourless solutions
- form HALIDE IONS with a charge of −1.

These trends are opposite to the ones in Group 1.

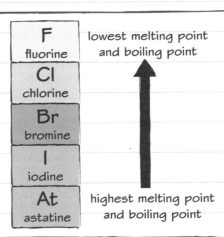

most reactive

least reactive

| F |
| fluorine |
| Cl |
| chlorine |
| Br |
| bromine |
| I |
| iodine |
| At |
| astatine |

lowest melting point and boiling point

highest melting point and boiling point

Displacement reactions

A more reactive halogen can DISPLACE a less reactive halogen from an aqueous solution of its salt. The table shows the results of adding halogens to solutions of halide ions.

Halogen		Halide ion in solution			Number of reactions
		chloride, Cl^-	bromide, Br^-	iodide, I^-	
	chlorine, Cl_2	✗	turns orange	turns brown	2
	bromine, Br_2	✗	✗	turns brown	1
	iodine, I_2	✗	✗	✗	0

For example:

Chlorine is more reactive than bromine so it displaces bromine from a solution of its salt:

chlorine + potassium bromide → potassium chloride + bromine

$$Cl_2(aq) + 2KBr(aq) \rightarrow 2KCl(aq) + Br_2(aq)$$

The displaced bromine causes the orange colour in the solution. Iodine causes the brown colour with iodides.

Explain the trend in reactivity down Group 7. *(3 marks)*

As you go down Group 7, the elements become less reactive. This is because as the atoms become larger, the outer shell becomes further from the nucleus so electrons are gained less easily in reactions.

As you go down Group 1, the atoms become larger and the outer shell becomes further from the nucleus. There is less attraction from the nucleus so the outer electron is lost more easily and the elements become more reactive.

Now try this

 1 State **two** properties that the Group 7 elements have in common.
(2 marks)

 2 Iodine is a solid at room temperature. Explain which state astatine will be found in at room temperature.
(2 marks)

3 (a) Describe what you **see** when bromine is added to potassium iodide solution.
(1 mark)

(b) Explain why this happens and write a word equation for the reaction. *(3 marks)*

 4 Explain why chlorine is less reactive than fluorine.
(3 marks)

 HIGHER

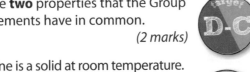

Hard and soft water

SOFT WATER easily forms lather when shaken with soap, but HARD WATER does not easily form lather.

Hard water

Hard water forms when CALCIUM IONS and MAGNESIUM IONS dissolve in it. This can happen in rivers, lakes and reservoirs when the water comes into contact with rocks containing magnesium compounds or calcium compounds.

Hard water reacts with soap to form a precipitate called **scum**. This means that more soap is needed to form lather. **Soapless detergents** do not form scum.

Types of hardness

There are two types of hard water:

- PERMANENT hard water stays hard when it is boiled.
- TEMPORARY hard water is softened when it is boiled.

Temporary hardness   HIGHER

Temporary hard water contains HYDROGENCARBONATE IONS, HCO_3^-. These decompose when the water is heated, producing CARBONATE IONS, CO_3^{2-}:

$$2HCO_3^-(aq) \xrightarrow{heat} CO_3^{2-}(aq) + CO_2(g) + H_2O(l)$$

The carbonate ions react with calcium ions and magnesium ions in the water to form PRECIPITATES. For example:

$$Ca^{2+}(aq) + CO_3^{2-}(aq) \rightarrow CaCO_3(s)$$

 Worked example   **D-C**

The volume of soap solution needed to form permanent lather is a measure of the hardness of water. It is found by **titration**.

Four samples of water were tested for hardness by shaking them with soap solution. The table shows the results.

Water sample (25 cm³)	Volume of soap solution needed to form lather in cm³
A	7
B	2
A after boiling	1
B after boiling	2

(a) Explain which water (A or B) is the hardest. *(2 marks)*

Water A is the hardest because it needed the most soap solution to form lather.

(b) Explain which water (A or B) contained temporary hardness. *(2 marks)*

Water A contained temporary hardness because it needed less soap after boiling but water B needed the same volume as before.

Now try this

 1 Describe how the experiment in the Worked Example could be made a fair test. *(2 marks)*

 2 Explain, with the help of an equation, how a precipitate can form without soap in temporary hard water that contains magnesium ions. *(5 marks)*

HIGHER

Softening hard water

Hard water can be softened by removing its dissolved calcium ions and magnesium ions.

Benefits and problems

✓ Hard water has calcium compounds that are good for teeth and bones, and help to reduce heart disease.

✗ Hard water increases costs because more soap is needed to form lather.

✗ Hard water increases costs because it reduces the efficiency of heating systems and kettles. When temporary hard water is heated, it can produce LIMESCALE. This coats the inside of pipes, boilers and kettles.

EXAM ALERT!

You might be asked to evaluate the use of a commercial water softener. You would need to balance the benefits of hard water against the problems it causes. Remember that water softeners also cost money to install and run.

Students have struggled with this topic in recent exams – **be prepared!**

Commercial water softeners such as **ion exchange columns** do the same job but are more expensive. They are in dishwashers, and may even be built into the water supply.

Washing soda

Sodium carbonate (washing soda) can soften hard water. It dissolves and reacts with:

• calcium ions to form a precipitate of calcium carbonate

• magnesium ions to form a precipitate of magnesium carbonate.

The water is softened because the dissolved calcium and magnesium ions are removed.

 Worked example   target **D-C**

Ion exchange columns also soften temporary hard water.

Use the diagram to describe how permanent hard water may be softened using an ion exchange column.

(3 marks)

The column is packed with beads made from an ion exchange resin. As the hard water flows through the column, its calcium ions are swapped for sodium ions from the resin. The water leaves without the calcium ions, so it is softened.

calcium ions displace sodium ions from the resin molecules

resin molecules — ion exchange column

calcium ions are retained by the resin molecules

Ion exchange resins also remove magnesium ions, which cause hardness too. Water leaving the column contains sodium ions, but these do not form scum with soap.

Now try this

target G-E   **1** State **one** benefit of hard water and **one** problem it causes.   (2 marks)

target D-C   **2** Describe how sodium carbonate softens hard water.   (2 marks)

target B-A*   **3** Dishwashers contain an ion exchange column. Suggest why dishwasher salt, pure sodium chloride, must be added every few weeks.   (3 marks)

Purifying water

Water is essential for life. Its quality is important.

Drinking water

Humans need drinking water with:

- ☑ low levels of microbes
- ☑ low levels of dissolved salts.

Fresh water can come from rivers, lakes or reservoirs. It must be:

- treated to remove solids such as leaves and sticks
- sterilised to reduce microbes that might cause disease.

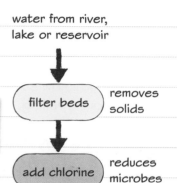

water from river, lake or reservoir

↓

filter beds — removes solids

↓

add chlorine — reduces microbes

↓

> The main stages in purifying water.

Distillation of water

Pure water can be obtained by DISTILLATION of seawater:

- seawater is boiled
- the water vapour is cooled and condensed to form pure water.

A lot of energy is needed to boil water, so distillation is rarely used in the UK and other countries where energy costs are high.

> Distillation is often used in hot countries. Heat from the sun can be used to evaporate water, rather than using heat from burning fuels.

> Water filters can be used to improve the taste and quality of tap water. They use carbon, silver, and ion exchange resins to remove dissolved substances.

Worked example D-C

Consider Fluoride may be added to drinking water. Give **two** reasons for doing this, and **two** arguments against doing this. *(4 marks)*

Fluoride strengthens teeth and reduces tooth decay. However, too much fluoride may cause weak bones and discoloured teeth. Some people do not like being forced to have fluoride in their drinking water.

> Other arguments against adding fluoride include:
>
> ☒ Fluoride can be obtained just by using fluoride toothpaste.
>
> ☒ Most water is used for bathing and cleaning, so the fluoride will cause unnecessary pollution.

Now try this

 **G-E**

 B-A*

1 State why filter beds and chlorine are needed to produce drinking water. *(2 marks)*

2 Silver particles and chlorine both sterilise water. Explain why water companies use chlorine but home filter jugs use silver. *(2 marks)*

3 Chlorine is a toxic gas. It can affect the taste and smell of drinking water. Evaluate the use of chlorine in drinking water. *(3 marks)*

Chemistry six mark question 1

There will be one six mark question on your exam paper, which will be marked for *quality of written communication* as well as scientific knowledge. This means that you need to apply your scientific knowledge, present your answer in a logical and organised way and make sure that your spelling, grammar and punctuation are as good as you can make them.

Worked example

Johann Döbereiner published his ideas about organising elements in 1829. He noticed that elements could be placed in atomic mass order into groups of three elements, called triads. Forty years later, Dmitri Mendeleev published his first periodic table, containing 60 elements.

Describe the similarities and differences between Döbereiner's triads, Mendeleev's table and the modern periodic table. *(6 marks)*

These are two of Döbereiner's five triads.

Alkali formers		Salt formers	
Li	7	Cl	35.5
Na	23	Br	80
K	39	I	127

Döbereiner's alkali formers and salt formers contain elements from Groups 1 and 7. The elements are in the same positions in their triads as they are in the modern groups. However, they are arranged in atomic mass order rather than in atomic number order, and they do not contain all the elements from these groups.

Mendeleev's table contained many more elements. If Döbereiner had five triads, he was only looking at fifteen elements. Mendeleev left space for elements that he thought would be discovered later, and he was able to successfully predict the properties of these missing elements.

The answer uses information from the diagram in the question and the periodic table, and scientific knowledge and understanding. It discusses two similarities and two differences.

EXAM ALERT!

You do not need to know about Döbereiner's work for the exam. However, you should be prepared to answer questions when given information, as in this question.

Students have struggled with exam questions similar to this – **be prepared!**

Data Sheet

In the exam, use the periodic table on the Data Sheet to help you answer questions.

Now try this

In the modern periodic table, the elements in each group have similar properties, and these properties gradually change as you go down a group.

Use your scientific knowledge and understanding to compare the properties of these elements from Group 1 and Group 7, and the trends in these two groups. *(6 marks)*

Group 1	Group 7
lithium	chlorine
sodium	bromine
potassium	iodine

Describe some of the main trends, chemical properties and physical properties of the elements in Group 1, and how they are different from the properties of the elements in Group 7.

Calorimetry

CALORIMETRY is a method to measure the amount of energy released or absorbed by a chemical reaction. Energy is normally measured in joules, J.

Energy from burning fuels

The energy released from a burning substance can be measured using a CALORIMETER. This is a container made from glass or metal.

To work out the amount of energy released, you have to measure:

• the volume of water in the calorimeter
• the change in temperature of the water.

You also need to know the SPECIFIC HEAT CAPACITY for water, which is 4.2 J/g/°C.

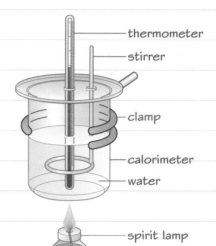
thermometer
stirrer
clamp
calorimeter
water
spirit lamp
ethanol

You can work out the energy released per gram of fuel if you measure the mass of fuel used in the experiment.

Energy from reactions in solution

The reactants are mixed in an insulated container when investigating energy changes in chemical reactions, such as solids reacting with water, or neutralisation reactions.

You will be given this equation in an exam:
$Q = mc\Delta T$
• Q = energy transferred, in J
• m = mass of water or solution, in g
• c = specific heat capacity of water
• ΔT = change in temperature, in °C

Worked example D-C

A student used a calorimeter to investigate the energy released from two fuels. In each case the temperature of 100 g of water increased by 20 °C. The table shows the results.

Fuel	Mass of fuel used in g
methanol	0.75
ethanol	0.60

(a) Calculate the energy released, in kilojoules, kJ, to heat the water. *(2 marks)*

$Q = mc\Delta T$
$Q = 100 \times 4.2 \times 20 = 8400$ J
8400 J $= 8400/1000 = 8.4$ kJ

(b) Calculate the energy released from each fuel in kJ/g *(4 marks)*

Energy from methanol = 8.4/0.75 = 11.2 kJ/g
Energy from ethanol = 8.4/0.60 = 14.0 kJ/g

Now try this

 D-C

1 25 g of sodium hydroxide solution was added to 25 g of hydrochloric acid. The temperature increased from 20 °C to 28 °C. Calculate the energy released in J. Assume that the specific heat capacity of the solution and the acid is the same as for water. *(2 marks)*

2 Bread contains 240 calories per gram. Calculate its energy content in J/g (1 calorie = 4.2 J). *(2 marks)*

Energy level diagrams

During a chemical reaction, energy is supplied to break bonds in the reactants, and energy is released when bonds are formed in the products. ENERGY LEVEL DIAGRAMS show the relative energies of the reactants and products.

Exothermic reactions

An exothermic reaction

Energy in kJ / Reaction time

reactants — energy given out to the surroundings — products

In an exothermic reaction:
- the energy of the reactants is more than the energy of the products.

Endothermic reactions

An endothermic reaction

Energy in kJ / Reaction time

products — energy taken in from the surroundings — reactants

In an endothermic reaction:
- the energy of the products is more than the energy of the reactants.

 Worked example 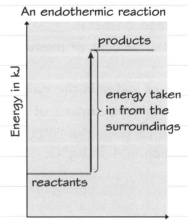 D-C

AQA SKILL Interpret Page 79

The energy level diagram represents an exothermic reaction.

(a) Explain what the energy change shown as A represents. *(2 marks)*

The activation energy, which means that this is the energy needed to start the reaction.

(b) Draw a line to represent the change in energy level when a catalyst is added. *(1 mark)*

energy ↑

A

B

reactants

C

products

In this diagram, C represents the overall energy change. A catalyst provides an alternative pathway with a lower activation energy for the chemical reaction.

 Now try this

D-C

1 Explain how you would know that the energy level diagram in the Worked Example shows an exothermic reaction. *(2 marks)*

2 Hydrogen burns in oxygen to form one product only.

    (a) Write a word equation for the reaction. *(1 mark)*

    (b) Explain why a flame is needed to start the reaction. *(1 mark)*

 B-A*

3 Referring to the Worked Example, explain why the line goes up then drops to a lower level. *(4 marks)*

Bond energies

HIGHER This whole page is Higher material.

The energy needed to break a bond is called its BOND ENERGY. The energy transferred in a reaction can be calculated from the bond energies involved.

Exothermic reactions	Endothermic reactions
Energy released when new bonds form is more than energy needed to break existing bonds	Energy needed to break existing bonds is more than energy released when new bonds form
energy in < energy out	energy in > energy out

Some bond energies

You will be given any bond energies you need in the exam question. The table shows some you may need on this page.

Bond	Bond energy in kJ
H–H	436
O=O	498
O–H	464
C–H	413
C=O	805

Hydrogen burns in oxygen to form water. The equation shows the structural formulae of the substances involved in this balanced reaction.

$$2 \times (H-H) + O=O \longrightarrow 2 \times \left(\underset{H}{} \overset{O}{} \underset{H}{} \right)$$

(a) Calculate the energy needed to break all the bonds in the reactants. *(3 marks)*

$2 \times (H–H) = 2 \times 436 = 872$ kJ

$1 \times (O=O) = 1 \times 498 = 498$ kJ

Total $= 872 + 498 = 1370$ kJ

(b) Calculate the energy released when new bonds form in the products. *(2 marks)*

$4 \times (O–H) = 4 \times 464 = 1856$ kJ

(c) Calculate the overall energy change for the reaction. *(2 marks)*

Energy change $= 1370 - 1856 = -486$ kJ

EXAM ALERT!

You will be given the structural formulae for the reactants and products in these questions. Use them to help you see how many of each type of bond there are.

Students have struggled with exam questions similar to this – **be prepared!**

Multiply the bond energy for each bond type by the number of those bonds present. Take care to use the correct values. For example, don't mix up O=O and O–O.

Each water molecule has two O–H bonds, so there are $2 \times 2 = 4$ O–H bonds in total.

The overall energy change is:
energy in − energy out
The negative sign shows that the reaction is exothermic and that energy is given out to the surroundings.

Now try this

Use the bond energies in the table at the top of the page to help you answer these questions.

B-A*

1 During electrolysis, water decomposes to form hydrogen and oxygen.

(a) Calculate the overall energy change in the reaction. *(4 marks)*

(b) Explain whether the process is exothermic or endothermic. *(2 marks)*

$$2 \times \left(\underset{H}{} \overset{O}{} \underset{H}{} \right) \longrightarrow 2 \times (H-H) + O=O$$

B-A*

2 Methane burns in oxygen to form carbon dioxide and water.

Calculate the overall energy change in the reaction. *(5 marks)*

$$H-\overset{\overset{H}{|}}{\underset{\underset{H}{|}}{C}}-H + 2 \times (O=O) \longrightarrow 2 \times \left(\underset{H}{} \overset{O}{} \underset{H}{} \right) + O=C=O$$
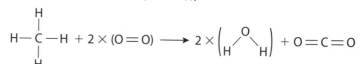

Hydrogen as a fuel

Internal combustion engines

Most vehicles have an INTERNAL COMBUSTION ENGINE. The fuel is burned inside the engine and the waste gases come out through the exhaust pipe. The fuel used is usually petrol or diesel, which contain HYDROCARBONS.

Hydrocarbon molecules are made up of hydrogen atoms and carbon atoms only.

Waste gases

Fuels produce waste substances, including:
- water vapour, H_2O
- carbon dioxide, CO_2
- carbon monoxide, CO
- sulfur dioxide, SO_2
- oxides of nitrogen, NO_x
- particulates (solid particles such as soot).

Burning hydrogen

Hydrogen can be burned as a fuel in internal combustion engines:

hydrogen + oxygen → water

$$2H_2(g) + O_2(g) \rightarrow 2H_2O(l)$$

Hydrogen can also be used as a fuel in FUEL CELLS. These produce electricity that can be used to drive an electric motor.

You do not need to know the details of the reactions that go on inside a fuel cell, but the overall reaction is: hydrogen + oxygen → water.

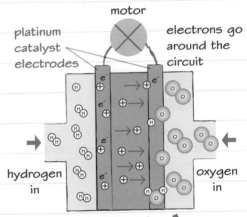

motor

platinum catalyst electrodes

electrons go around the circuit

hydrogen in

oxygen in

hydrogen ions (protons) cross the membrane

water out

Hydrogen can be used as a fuel for internal combustion engines (ICE) and for fuel cells (FC). Use information from the table to evaluate its use in these two ways. *(4 marks)*

Property	ICE	FC
technology	established	under development
efficiency	35%	45%
noise	noisy	quiet
nitrogen oxides (NOx) produced	yes	no
system cost	low	high

When used for internal combustion engines rather than for a fuel cell, hydrogen is less efficient and produces NO_x. However, when used for fuel cells, the costs of the system are high and the technology is still under development. Overall I think fuel cells are better because of their high efficiency, quietness and no NO_x produced.

Now try this

 target G-E

1 Give the word equation for the combustion of hydrogen. *(1 mark)*

 target D-C

2 (a) Vehicles powered by fuel cells are very quiet. Suggest **one** advantage and **one** disadvantage of this property. *(2 marks)*

(b) Suggest a reason why fuel cells are expensive. *(1 mark)*

 target B-A*

3 Give **two** advantages and **two** disadvantages to using hydrogen, rather than hydrocarbons, as a fuel in internal combustion engines. *(4 marks)*

Tests for metal ions

Metal ions can be identified using simple laboratory tests.

Flame tests

Some metal ions can be identified using FLAME TESTS. Different metal ions in compounds produce different and distinctive colours in flame tests.

Metal ion	Flame test colour
lithium, Li^+	crimson
sodium, Na^+	yellow
potassium, K^+	lilac
calcium, Ca^{2+}	red
barium, Ba^{2+}	green

Clean the flame test loop in acid each time, rinse with water and check it is clean in the Bunsen burner flame.

To test a substance, dip the clean loop in a solution of the ions and hold at the edge of a blue flame.

Hydroxide precipitates

Some metal ions form coloured hydroxide PRECIPITATES. The sample solution is placed in a test tube and a few drops of dilute sodium hydroxide solution are added. The table shows the colours you need to know.

Metal ion	Colour of precipitate
copper(II), Cu^{2+}	blue
iron(II), Fe^{2+}	green
iron(III), Fe^{3+}	brown

Some metal ions form white precipitates with sodium hydroxide solution:
- magnesium ions, Mg^{2+}
- calcium ions, Ca^{2+}
- aluminium ions, Al^{3+}.

Copper and iron are **transition metals**. Many of these elements form coloured compounds. Although sodium hydroxide is soluble in water, most hydroxides are insoluble and so form precipitates in these tests.

Worked example

D-C

Describe how to distinguish between the hydroxide precipitate formed by aluminium ions and the one formed by magnesium ions. *(2 marks)*

The aluminium hydroxide precipitate will dissolve in excess sodium hydroxide but the magnesium hydroxide precipitate will not.

Calcium ions also form a hydroxide precipitate that does not dissolve in excess sodium hydroxide solution.

Now try this

G-E 1 Describe a laboratory test to distinguish between iron(II) chloride solution and iron(III) chloride solution. *(3 marks)*

D-C 2 Calcium in the diet is important for the development of healthy bones and teeth.

(a) Describe **two** ways in which calcium may be detected. *(4 marks)*

(b) Explain why it is not possible to distinguish magnesium from calcium in one of these tests. *(2 marks)*

D-C 3 Suggest why it is difficult to identify a mixture of ions in a flame test. *(1 mark)*

More tests for ions

Non-metal ions can be identified using simple laboratory tests. When combined with tests for metal ions, unknown compounds can be identified.

Tests for carbonate, sulphate and halide ions

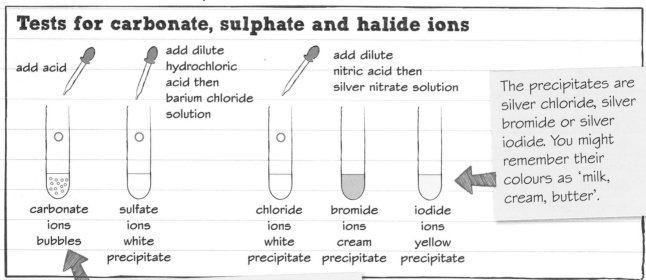

add acid

add dilute hydrochloric acid then barium chloride solution

add dilute nitric acid then silver nitrate solution

The precipitates are silver chloride, silver bromide or silver iodide. You might remember their colours as 'milk, cream, butter'.

| carbonate ions bubbles | sulfate ions white precipitate | chloride ions white precipitate | bromide ions cream precipitate | iodide ions yellow precipitate |

The bubbles are carbon dioxide. This gas reacts with limewater to produce a white precipitate.

Worked example   D–C

AQA SKILL Interpret Page 79

A white powder is used to treat water for drinking, and wastewater before it is discharged into rivers.

It produces a white precipitate with sodium hydroxide solution, which dissolves in excess alkali. It does not bubble when dilute hydrochloric acid is added to it, but produces a white precipitate when barium chloride is then added. Explain what the name and formula of the powder is.   *(4 marks)*

The test with sodium hydroxide solution shows that the powder contains aluminium ions. The test with barium chloride shows that it contains sulfate ions. It must have been aluminium sulfate, $Al_2(SO_4)_3$.

Now try this

G–E

1 A student tested some washing soda with dilute hydrochloric acid to show that it is a carbonate. Carbon dioxide gas was given off in the test.

(a) Describe what you would see when the carbon dioxide is given off.   *(1 mark)*

(b) Describe how you could show that the gas is carbon dioxide.   *(2 marks)*

(c) Washing soda gives a yellow flame test result. Explain what the chemical name is for washing soda.   *(2 marks)*

D–C

2 A cargo of clothing was damaged by water while at sea. It could have been caused by rainwater or by seawater. The table shows the results of tests using silver nitrate solution. Explain what the tests show.   *(3 marks)*

Sample	Test result
Rainwater	No visible change
Seawater	White precipitate
Damp from clothing	White precipitate

3 The labels have come off four bottles. The bottles contain aluminium sulfate, magnesium sulfate, potassium iodide or potassium bromide. Describe how you would use chemical tests to identify the contents of each bottle.   *(6 marks)*

B–A*

You will also need to refer to tests on the previous page.

Titration

HIGHER This whole page is Higher material.

TITRATION is used to measure the volumes of acid and alkali that react with each other.

Doing a titration

A typical titration goes like this:
- acid is put into a burette
- alkali is put in a flask using a pipette
- a suitable indicator is added to the alkali
- the start reading on the burette is noted
- acid is added until the colour changes (the END-POINT)
- the end reading is noted.

Precautions to work safely and accurately include:

✓ use a pipette filler for the pipette

✓ put the flask onto a white tile

✓ swirl the flask to mix the liquids

✓ wear eye protection

✓ repeat the experiment until you get consistent results.

Worked example

In a titration, 25.0 cm³ of 0.50 mol/dm³ potassium hydroxide was added to a flask. A few drops of phenolphthalein indicator were added. The colour changed when 31.5 cm³ of hydrochloric acid was added from a burette.

(a) Complete the balanced equation for the reaction. *(1 mark)*

$HCl + KOH \rightarrow KCl + H_2O$

(b) Calculate the number of moles of potassium hydroxide used. *(2 marks)*

volume of potassium hydroxide
= 25.0 cm³ ÷ 1000 = 0.025 dm³

moles = concentration × volume
= 0.50 mol/dm³ × 0.025 dm³ = 0.0125 mol

(c) Explain the number of moles of acid used. *(2 marks)*

The balanced equation shows that one mole of HCl reacts with one mole of KOH. If the number of moles of acid and alkali are the same, there must be 0.0125 mol of acid.

(d) Use your answer to part **(c)** to calculate the concentration of the hydrochloric acid. *(2 marks)*

volume of hydrochloric acid
= 31.5 cm³ ÷ 1000 = 0.0315 dm³

concentration = moles ÷ volume
= 0.0125 mol ÷ 0.0315 dm³
= 0.40 mol/dm³

Now try this

1   25.0 cm³ of 0.20 mol/dm³ potassium hydroxide was added to a flask. The volume of hydrochloric acid needed to neutralise it was 20.0 cm³. Calculate the concentration of the acid. *(4 marks)*

2   25.0 cm³ of sodium hydroxide solution was added to a flask. The volume of 0.50 mol/dm³ hydrochloric acid needed to neutralise it was 24.0 cm³.

$HCl + NaOH \rightarrow NaCl + H_2O$

(a) Calculate the concentration of the sodium hydroxide solution in mol/dm³. *(4 marks)*

(b) The relative formula mass, M_r of NaOH is 40.
Calculate the concentration of the solution in g/dm³. *(1 mark)*

mass = M_r × mol

Chemistry six mark question 2

There will be one six mark question on your exam paper, which will be marked for *quality of written communication* as well as scientific knowledge. This means that you need to apply your scientific knowledge, present your answer in a logical and organised way and make sure that your spelling, grammar and punctuation are as good as you can make them.

Worked example

AQA SKILL
Evaluate
Page 79

Petrol is commonly used as a fuel for cars. Hydrogen may also be used but different, more expensive engines are needed. The table shows some differences between the two fuels.

Use this information, and your knowledge and understanding, to evaluate the use of hydrogen as a fuel for cars compared to petrol. *(6 marks)*

	Petrol	Hydrogen
Cost in pence/litre	135	55
Energy content in MJ/litre	34	8.5
UK filling stations selling the fuel in 2012	8480	1

Petrol costs almost three times more per litre than hydrogen, but it contains four times more energy per litre. This makes petrol a more cost-effective fuel than hydrogen. Also, there are many more petrol stations than hydrogen stations, so petrol is a more practical fuel.

However, petrol comes from crude oil, a non-renewable resource. Hydrogen can be produced by the electrolysis of water using renewable resources such as wind power.

Petrol produces carbon dioxide when it burns, but hydrogen only produces water. As carbon dioxide is a greenhouse gas linked to global warming, hydrogen fuel has less impact on the environment.

Use the data

Remember to use the information given and explain what it means, rather than just repeating it.

The answer uses information from the table to compare petrol favourably with hydrogen.

Calculations could be made to support the answer. The energy cost of petrol is 135/34 = 4p/MJ, and for hydrogen it is 55/8.5 = 6.5p/MJ.

The use of 'however' is a good way to indicate that the answer is moving to a different point of view. Scientific knowledge and understanding is used to discuss the advantages of hydrogen.

Now try this

Hard water contains dissolved calcium or magnesium compounds. There are certain health benefits to hard water, but it can also cause problems. For example, limescale forms in heating systems and kettles. The table shows its typical effects on a central heating boiler.

Limescale thickness in mm	% increase in energy use
2	15
4	30
6	43
8	55

Use the information given, and your scientific knowledge and understanding, to discuss the environmental, social and economic impacts of hardness in water.

(6 marks)

The Haber process

AMMONIA is used to make ammonium salts for fertilisers. It is made in the HABER PROCESS.

Raw materials

The raw materials for the Haber process are nitrogen and hydrogen:

- NITROGEN is obtained from the air
- HYDROGEN is obtained from natural gas or other sources.

Hydrogen is also produced by the electrolysis of sodium chloride solution.

The reaction is reversible

nitrogen + hydrogen ⇌ ammonia

$$N_2(g) \ + \ 3H_2(g) \ \rightleftharpoons \ 2NH_3(g)$$

⊗ Only some of the nitrogen and hydrogen react together to produce ammonia.

⊗ Ammonia breaks down again into nitrogen and hydrogen.

✓ Unreacted nitrogen and hydrogen are recycled.

The Haber process

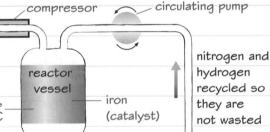

compressor   circulating pump

nitrogen from air

hydrogen

200 atmospheres pressure, temperature 450 °C

reactor vessel

iron (catalyst)

nitrogen and hydrogen recycled so they are not wasted

A diagram to show the Haber process.

gas is

cooled   liquid ammonia continuously removed

Worked example   **D-C**

The graph shows how the yield of ammonia varies as the temperature and pressure are changed.

Give **two** reasons why a higher temperature than 450 °C may not be desirable.   **(2 marks)**

The yield of ammonia is lower at higher temperatures.

More energy will be needed to heat the reaction to a higher temperature than 450 °C.

Increased energy use has an environmental impact too. For example, more carbon dioxide may be released from burning fuels.

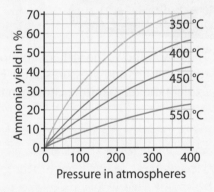

Now try this

Refer to the diagram and graph above when answering these questions.

G-E

1 (a) What is the percentage yield of ammonia at 200 atmospheres and 450 °C?   *(1 mark)*

    (b) Explain how ammonia is removed from the unreacted nitrogen and hydrogen.   *(2 marks)*

D-C

2 (a) Explain why iron is needed in the Haber process.   *(2 marks)*

    (b) Explain why nitrogen and hydrogen must be recycled in the Haber process.   *(2 marks)*

B-A*

3 (a) Balance this equation for the Haber process: $N_2 + \ldots H_2 \rightleftharpoons \ldots NH_3$.   *(1 mark)*

    (b) Explain how the yield of ammonia may be increased in the Haber process.   *(2 marks)*

Equilibrium

HIGHER This whole page is Higher material.

Equilibrium

When a reversible reaction happens in a CLOSED SYSTEM, a situation called EQUILIBRIUM is reached. At equilibrium:

- the forward and backward reactions still happen
- both reactions happen at exactly the same rate.

Equilibrium amounts

The relative amounts of all the reacting substances at equilibrium depend on the conditions of the reaction.

The OPTIMUM CONDITIONS in an industrial process are the conditions that allow the highest yield for the least money. Often the optimum conditions depend on getting the temperature and pressure right.

Changing the pressure

In a reversible reaction involving gases, the equilibrium moves in the direction of the fewest molecules of gas.

For example, sulfur dioxide reacts with oxygen in a reversible reaction:

$$2SO_2(g) + O_2(g) \rightleftharpoons 2SO_3(g)$$

Increasing the pressure favours the forward reaction, so more SO_3 is present at equilibrium.

Changing the temperature

When the temperature is INCREASED, the equilibrium moves in the direction of the ENDOTHERMIC process.

For example, carbon monoxide reacts with hydrogen to produce methanol:

$$CO(g) + 2H_2(g) \rightleftharpoons CH_3OH(l)$$

The forward reaction is exothermic. This means:

- the backward reaction is endothermic
- the yield of methanol is reduced at higher temperatures.

Worked example

The graph shows the yield of ammonia at different temperatures and pressures in the Haber process. Explain the temperature chosen in the Haber process.

(4 marks)

The yield of ammonia increases as the temperature is decreased, showing that the equilibrium is moved to the left. However, at lower temperatures the rate of reaction will be lower. This would make the process uneconomic, so an optimum temperature of about 450 °C is used.

Note that building equipment to withstand very high pressures is expensive and increases the risk to safety.

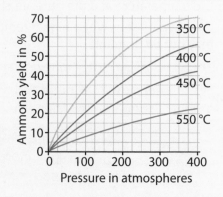

Now try this

Refer to the graph when answering these questions.

1 (a) State what happens to the yield of ammonia as the pressure is increased. *(1 mark)*

(b) Explain what will happen to the rate of reaction as the pressure is increased. *(2 marks)*

2 Nitrogen reacts with hydrogen in the Haber process: $N_2(g) + 3H_2(g) \rightleftharpoons 2NH_3(g)$.

(a) Describe what happens when a reaction reaches equilibrium. *(2 marks)*

(b) Explain the pressure chosen in the Haber process. *(4 marks)*

Alcohols

ALCOHOLS are used as solvents and fuels. ETHANOL is the main alcohol in alcoholic drinks such as wine and beer.

Structures

Alcohols all contain the FUNCTIONAL GROUP –OH.

methanol

Names	Structure
End in ol.	Has a hydroxyl group.

ethanol

The alcohols form a HOMOLOGOUS SERIES. They have:

- the same functional group
- similar chemical properties.

You should be able to recognise alcohols from their names or formulae:

- methanol is CH_3OH
- ethanol is C_2H_5OH
- propanol is C_3H_7OH.

Reactions of alcohols

Methanol, ethanol and propanol all:

- dissolve in water to form a neutral solution
- react with sodium to produce hydrogen
- burn in air.

For example:

methanol + oxygen → carbon dioxide + water

$2CH_3OH(l) + 3O_2(g) \rightarrow 2CO_2(g) + 4H_2O(l)$

Oxidation of ethanol

Burning, or combustion, is an example of an oxidation reaction. Ethanol burns in air:

ethanol + oxygen → carbon dioxide + water

$C_2H_5OH(l) + 3O_2(g) \rightarrow 2CO_2(g) + 3H_2O(l)$

Ethanol can also be oxidised to ETHANOIC ACID by:

- chemicals called OXIDISING AGENTS, or
- the action of microbes.

Ethanoic acid is the main acid in vinegar.

Worked example

 D-C

The diagram shows the structure of propanol.

Explain how you know from its structure that propanol is an alcohol and not an alkane. *(2 marks)*

Alkanes are hydrocarbons, so they only contain hydrogen and carbon atoms, but propanol contains an oxygen atom in its –OH group.

Now try this

1 Which statement about ethanol is correct? Tick (✓) **one** box. *(1 mark)*

Statement	Tick (✓)
Ethanol burns in air to form ethanoic acid.	
Ethanol reacts with sodium to produce hydrogen.	
Ethanol dissolves in water to produce an acidic solution.	

 D-C

2 (a) State **two** ways you know that hexanol, $C_6H_{13}OH$, is an alcohol. *(2 marks)*

(b) Suggest **one** use for hexanol. *(1 mark)*

 B-A*

3 Balance this equation for the combustion of propanol:

$2C_3H_7OH + 9O_2 \rightarrow \ldots CO_2 + \ldots$ *(2 marks)*

Carboxylic acids

Vinegar is an aqueous solution of ethanoic acid, an example of a carboxylic acid.

Structures

Carboxylic acids contain the FUNCTIONAL GROUP –COOH.

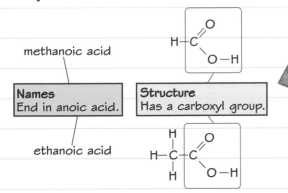

methanoic acid

Names	Structure
End in anoic acid.	Has a carboxyl group.

ethanoic acid

You should be able to recognise carboxylic acids from their names or formulae:
- methanoic acid is HCOOH
- ethanoic acid is CH_3COOH
- propanoic acid is C_2H_5COOH.

Acids make solutions acidic because they release hydrogen ions, $H^+(aq)$.

Reactions of carboxylic acids

Carboxylic acids:

- dissolve in water to form acidic solutions
- react with carbonates to produce carbon dioxide
- react with alcohols to produce ESTERS when an acid catalyst is added.

Acidity of carboxylic acids HIGHER

Carboxylic acids are WEAK ACIDS because they do not ionise completely when they are dissolved in water:

$$CH_3COOH(aq) \rightleftharpoons CH_3COO^-(aq) + H^+(aq)$$

At the same concentration:
- carboxylic acids have a higher pH value than strong acids such as hydrochloric acid.

The diagram shows the structure of propanoic acid.

Describe **two** tests you could do to show that propanoic acid solution is acidic. *(4 marks)*

1 Add universal indicator. The mixture should go red or orange, showing that it is acidic.

2 Add it to some calcium carbonate and look for fizzing. The gas given off should turn limewater cloudy.

 Now try this

1 Which statement about ethanoic acid is correct? Tick (✓) **one** box. *(1 mark)*

target G-E

Statement	Tick (✓)
Ethanoic acid dissolves in water to produce a neutral solution.	
Ethanoic acid reacts with carbonates to produce carbon dioxide.	
Ethanoic acid reacts with esters to produce alcohols.	

2 (a) Explain how you know that butanoic acid, C_3H_7COOH, is a carboxylic acid. *(2 marks)*

target D-C

(b) Name the type of substance formed when butanoic acid reacts with ethanol. *(1 mark)*

3 Explain why 0.5 mol/dm³ hydrochloric acid has a pH of 0.3 but ethanoic acid at the same concentration has a pH of 2.5. *(4 marks)*

 target B-A*

HIGHER

Esters

ESTERS are volatile compounds (they evaporate easily). They have distinctive smells and are used as perfumes and food flavourings.

Structures

Esters contain the FUNCTIONAL GROUP –COO–.

ethyl
ethanoate

Structure
Has a ester group.

You should be able to recognise esters from their names or formulae.

- In general, their names are like this:

  ....yl ....anoate.
- Their formulae contain the –COO– group.

  For example, the formula for ethyl ethanoate is $CH_3COOC_2H_5$.

Making ethyl ethanoate

Ethyl ethanoate is made from ethanoic acid and ethanol:

ethanoic acid + ethanol → ethyl ethanoate + water

$CH_3COOH(aq) + C_2H_5OH(aq) → CH_3COOC_2H_5(aq) + H_2O(l)$

ethanoic acid   ethanol   ethyl ethanoate   water

Worked example D-C

Aspirin is an ester used as a medicine to reduce pain and fever. In the past it was extracted from the bark of willow trees. Nowadays it is manufactured. One way to make aspirin is to react salicylic acid with ethanoic acid.

(a) Suggest how the rate of the reaction could be increased. *(2 marks)*

Add sulfuric acid to act as a catalyst.

(b) Suggest **two** advantages of making aspirin rather than getting it from natural sources. *(2 marks)*

Large amounts can be made as needed, and it can be more easily purified.

Now try this

1 Esters are used in perfumes. Give **one** other use of esters. *(1 mark)*

2 (a) Give the formula of the functional group that esters have in common. *(1 mark)*

  (b) Name the ester formed when ethanol reacts with ethanoic acid. *(1 mark)*

  (c) Suggest why hydrochloric acid may also be added to the reaction mixture. *(2 marks)*

3 Explain how the properties of esters make them suitable for use as perfumes. *(3 marks)*

4 The diagram shows the structure of methyl butanoate, an ester that smells of apple. State, with the help of the diagram, why methyl butanoate is an ester. *(2 marks)*

Using organic chemicals

Alcohols, carboxylic acids and esters are ORGANIC CHEMICALS. There are social and economic advantages and disadvantages in their use.

Uses of organic chemicals

Alcohols are used for:
- fuels
- solvents.

Esters are used for:
- food flavourings
- perfumes.

> Ethanol is the alcohol in alcoholic drinks such as beer and wine. It is a legal recreational drug but it can cause damage to the body.
>
> Ethanol can be oxidised to ethanoic acid, a carboxylic acid. This is the main acid in vinegar.

Worked example   D-C

Chemical manufacturers must produce an MSDS (Material Safety Data Sheet) for each substance they make. The MSDS describes the hazards associated with a particular substance. Part of an MSDS for concentrated ethanoic acid is shown on the right.

(a) Explain why it might be dangerous to manufacture concentrated ethanoic acid on a large scale.   *(2 marks)*

Ethanoic acid could harm people in the factory because it can cause skin burns, and it could be set on fire or explode if it gets too hot.

(b) Vinegar is a dilute solution of ethanoic acid. Evaluate the use of vinegar in food.
    (3 marks)

Vinegar is used to flavour food. Concentrated ethanoic acid may be dangerous, but it is diluted in vinegar so it is safe to use. Overall I think it is worth having vinegar for food.

Ethanoic acid
Potential Health Effects:
Very hazardous in contact with skin or eyes, or when swallowed or breathed in. Skin contact may produce burns, reddening or blistering.
If spilled:
Mop up with plenty of water. Keep away from heat and flames.
Flammability:
Flammable. Can form explosive mixtures in air.

> Remember that very concentrated chemicals are usually much more dangerous than the same chemical in small amounts or when diluted.

> Higher Tier students could also describe ethanoic acid as a weak acid, because carboxylic acids do not ionise completely when dissolved in water.

Now try this

1 Read this information about fruit flavours, then answer the question.

> Natural fruit flavours often consist of several different substances. Together they make the distinctive tastes and smell of fruits. It can be difficult and expensive to extract natural flavours to use in food. Esters are used as artificial food flavourings. They are cheap to manufacture but usually a few different substances are mixed to make the flavouring.

Evaluate the use of esters as artificial flavours.
Give **one** advantage and **one** disadvantage.   *(3 marks)*

2 Ethanol may be used as a fuel on its own or mixed with other fuels such as petrol. It can be made by fermenting food crops such as sugar cane, wheat or maize. Use your scientific knowledge and understanding to evaluate the use of ethanol produced in this way as a fuel.   *(4 marks)*

Chemistry six mark question 3

There will be one six mark question on your exam paper, which will be marked for *quality of written communication* as well as scientific knowledge. This means that you need to apply your scientific knowledge, present your answer in a logical and organised way and make sure that your spelling, grammar and punctuation are as good as you can make them.

Worked example

Common table salt is sodium chloride. Manufacturers may add other ingredients to their product. For example, calcium carbonate may be added to help it flow freely. The health benefits of table salt are improved in 'fortified' salts by adding a little iron fumarate or potassium iodide, while 'low-salt' products are mostly potassium chloride.

Four salt samples (A, B, C and D) were tested in a public health laboratory. The results are shown in the table. Use your knowledge and understanding to interpret the results. *(6 marks)*

Salt	Sodium hydroxide added	Flame test colour	Nitric acid added then silver nitrate
A	faint white precipitate	yellow	bubbles, white precipitate
B	no change	lilac	white precipitate
C	no change	yellow	yellow precipitate
D	green precipitate	yellow	white precipitate

The flame tests show that all the salts, except Salt B, contain sodium ions. The flame test result for Salt B shows that it contains potassium ions. The silver nitrate test shows that it contains chloride ions, so B must be a 'low-salt' product.

The silver nitrate test for Salt C shows that it contains iodide ions, so it must be a 'fortified' salt containing potassium iodide.

The green precipitate from Salt D shows that it contains iron(II) ions, so it must be a 'fortified' salt containing iron(II) fumarate.

Salt A releases bubbles when acid is added, so it contains carbonate ions. It must contain calcium carbonate to help it flow, and the sodium hydroxide test confirms this.

The lilac colour from the potassium ions is hidden by the yellow colour from the sodium ions in the flame test for Salt C. This is because only a small amount of the potassium salt is added to the sodium salt.

The answer uses information from the table to explain which ions must be present in the salt samples. It considers the flame test results first, but the other tests could have been discussed first.

Iron(III) ions would produce a brown precipitate instead, so the full identity of the iron compound is given.

A white precipitate with sodium hydroxide could also indicate aluminium or magnesium ions, but the information in the question only mentions **calcium carbonate**.

Now try this

Some oven cleaners contain sodium hydroxide. As this is a strong alkali, the cleaner is safer when diluted with water, but users still need to wear gloves. The concentration of sodium hydroxide can be found if the volume of acid needed to neutralise the alkali is measured.

Describe how the volume of hydrochloric acid that reacts with a known volume of diluted oven cleaner can be found by titration. Include the names of the apparatus and substances needed, and a brief risk assessment to identify hazards and precautions needed. *(6 marks)*

X-rays

X-rays, like light, are part of the ELECTROMAGNETIC SPECTRUM.

X-rays are electromagnetic waves with a very short wavelength. Their wavelength is about the same as the diameter of an atom. X-rays are a type of IONISING RADIATION. They affect photographic film in a similar way to visible light.

'Ionising' means that atoms gain or lose electrons and become charged. When molecules in the body become ionised, they can cause cancer.

Using X-rays safely

Too much exposure to X-rays can cause cancers or death. Safety precautions include:
- warning lights to show when X-rays are being produced
- shielding the X-ray source with lead
- keeping X-ray operators away from the X-ray source when it is operating (usually they go behind a shield or into a separate room)
- wearing goggles and protective clothing such as a lead apron.

Worked example

1 The photograph shows an X-ray image of a patient with an artificial hip joint.

Explain why X-rays are useful to investigate bone fractures or problems with artificial joints, but are not very good for investigating problems with soft tissues such as muscles. *(3 marks)*

X-rays hitting the photographic film make it go black. If an absorbing material such as bone or metal is in the way, that area of the film stays white.

Soft tissues such as muscles transmit some of the X-rays, so they only show up faintly on the image.

Transmit means 'allows rays to pass through'.

Worked example B-A*

2 Explain why a person with a suspected broken leg will have an X-ray photograph taken, but the operator will stand behind a screen while the photograph is being taken. *(2 marks)*

X-rays can cause cancers. The risk of getting cancer during the taking of one X-ray photograph is small. An X-ray operator will take a large number of X-rays in a year and so must reduce his or her exposure by using the screen to block the X-rays.

Now try this

1 Look at the X-ray photograph shown above. Explain how the photograph shows that the patient has a metallic artificial hip bone. *(2 marks)*

2 Shoe shops used to have X-ray machines to show shoppers the bones in their feet inside new shoes. Dentists regularly take X-ray photographs of patients' teeth. Explain why dentists still use X-ray machines but shoe shops do not. *(3 marks)*

Ultrasound

ULTRASOUND has a frequency higher than the upper limit of hearing for humans. Ultrasound can be produced by electronic systems. The human ear can detect sounds in the range 20–20 000 Hz.

Reflection of ultrasound

Ultrasound waves are partially reflected at the boundary between two different MEDIA (materials). For example, if a fetus in the womb is scanned with ultrasound the waves are partially reflected by the boundary between the fetus and the fluid surrounding it.

Ultrasound equation

The distance travelled by an ultrasound wave is given by the equation:

$s = v \times t$

- s is the distance travelled by the sound in metres, m
- v is the velocity of the sound in metres per second, m/s
- t is the time taken in seconds, s

Ultrasound scans

A probe sends out pulses of ultrasound and detects the echoes. A computer converts the time taken for the echoes to arrive into an image. The image can then be used to check the growth and development of the fetus.

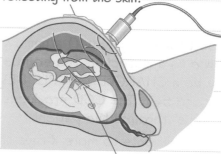

A gel is used to stop the ultrasound just reflecting from the skin.

Some sound is reflected when the ultasound waves pass into a different medium, such as fat or bone.

Worked example B-A*

The oscilloscope screen shows the reflections of an ultrasound pulse travelling into the womb of a pregnant woman. There are reflections at the wall of the womb and the skin of the fetus. Calculate the distance between the fetus and the wall of the womb. The speed of sound in the body = 1500 m/s.   *(4 marks)*

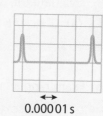

↔ 0.000 01 s

time difference between echoes = 0.000 01 s × 4 = 0.000 04 s.

Distance travelled = 1500 m/s × 0.000 04 s = 0.06 m

Distance between fetus and wall = ½ × 0.06 m = 0.03 m.

When ultrasound is reflected it goes there and back, so you have to halve the distance travelled.

Now try this

 D-C

1 Of the frequencies of sounds below:

A: 5 Hz   B: 500 Hz
C: 5000 Hz   D: 500 000 Hz

(a) which **two** can be heard by most humans?   *(2 marks)*

(b) which is ultrasound?   *(1 mark)*

 B-A*

2 A probe sends a pulse of ultrasound from a patient's skin through his stomach. One echo arrives after 0.000 05 s and a second echo after 0.000 21 s.

(a) Explain why two echoes are produced.   *(2 marks)*

(b) Calculate the diameter of the patient's stomach.   *(4 marks)*

Medical physics

Diagnosis and treatment

- X-ray photographs are used to detect broken bones or cavities in teeth.
- X-ray CT scans give a more detailed picture of organs in the body.
- X-rays are used to kill cancer cells.
- Ultrasound is used to shake kidney stones into tiny fragments that can be excreted.

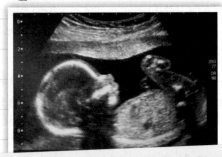

Ultrasound scans are used to examine a foetus in the mother's womb and to look for and destroy stones in the kidney and gall bladder.

CT scans

In computerised tomography (CT) a narrow beam of X-rays passes through the patient from the source to a detector. The source and detector rotate around the patient. The detector may be a CHARGED COUPLE DEVICE (CCD) similar to the ones used in a digital camera (see page 56). The CCD produces a signal that the computer then builds up into a picture of a 'slice' through the patient.

CT scans expose patients to more X-rays than a single picture, giving a higher risk of cancer. A CT scan is used if the benefits outweigh the risk. Few are carried out on a patient.

Worked example

B-A*

AQA SKILL
Evaluate
Page 79

Evaluate the use of X-rays, CT scans and ultrasound in the following medical situations:

(a) Checking the development of a foetus.   *(3 marks)*

It would be best to use an ultrasound scan because ultrasound is not harmful. The quality of the image is not high but it is high enough to see the detail necessary.

(b) Examining a broken arm bone.   *(3 marks)*

An X-ray photograph would give a clear picture of the broken bone. X-rays are harmful but the dose from one photograph would be a very low risk.

(c) Detecting a tumour in the liver.   *(3 marks)*

A CT scan would give a very detailed picture of the liver, which ultrasound or an X-ray photograph could not provide. The X-rays could be harmful but the benefit of locating the tumour outweighs the risks.

Now try this

D-C

1 Hospitals can use X-ray photographs, CT scans and ultrasound waves to diagnose problems. Which technique would be used in each of these cases?

   (a) Looking for a tumour in the brain.   *(1 mark)*

   (b) Breaking up kidney stones.   *(1 mark)*

   (c) Checking fillings in teeth.   *(1 mark)*

   (d) Checking the health of a foetus.   *(1 mark)*

B-A*

2 As many as 1% of new cancers may be caused by CT scans.

   (a) State **two** benefits of CT scans.   *(2 marks)*

   (b) Evaluate the use of ultrasound as an alternative to CT scans.   *(4 marks)*

Refraction in lenses

Refraction

Light changes direction when it passes from one medium to another. This is called REFRACTION. The size of the change in direction is determined by the REFRACTIVE INDEX of the medium the light enters.

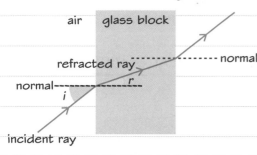

air   glass block

refracted ray ---------- normal

normal--------
i r

incident ray

Refractive index

Refractive index $= \dfrac{\sin i}{\sin r}$

- i is the angle of incidence
- r is the angle of refraction

> The normal is the line at 90° to a plane surface.

> In this diagram the media are air and glass.

Worked example

 B-A*

A spectacle lens is made from a transparent plastic. A ray of light enters the lens with an angle of incidence of 60°. The angle of refraction is 30°. Calculate the refractive index of the plastic.   *(2 marks)*

refractive index $= \dfrac{\sin 60°}{\sin 30°}$

$= \dfrac{0.866}{0.5} = 1.7$

The answer is to 2 significant figures, like the data given.

EXAM ALERT!

When calculating the refractive index, remember it is the sine of the angle, not the angle itself, that is needed. Make sure you know how to use your calculator to find the 'sin' of angles.

> Students have struggled with exam questions similar to this – **be prepared!**

Lenses

Light passing through a lens is refracted and forms an image. In a CONVEX or CONVERGING LENS, rays of light parallel to the axis bend so that they pass through a point called the PRINCIPAL FOCUS of the lens.

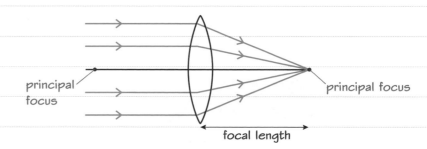

principal focus

principal focus

focal length

Now try this

> You need to use sin⁻¹ or 'inverse sin' to work out the angle.

 D-C

1 If a convex lens is used to direct sunlight onto a piece of paper the paper may catch fire. Explain why this happens.   *(2 marks)*

 B-A*

2 Diamond has the highest refractive index known of 2.42. Calculate the angle of refraction when a ray of light hits the surface of a diamond with an angle of incidence of 30°.   *(3 marks)*

Images and ray diagrams

Lenses can form real or virtual images.

A REAL IMAGE is an image that can be displayed on a screen. A lens forms a real image when it makes rays of light converge.

A VIRTUAL IMAGE cannot be displayed on a screen. You see a virtual image when a lens bends light so that it seems to come from a point behind the lens.

The OBJECT is the thing you are looking at.

Describing images

An image can be described by:

- its size relative to the object – if it is larger it is MAGNIFIED and if it is smaller it is DIMINISHED
- whether it is UPRIGHT or INVERTED (upside down)
- whether it is REAL or VIRTUAL.

Ray diagrams for convex lenses

Convex lenses are thicker in the middle than at the edges. They can produce real or virtual images. You can work out the type of image formed by a lens by drawing a ray diagram.

A convex lens can be shown as ↕.

A: Draw a vertical line representing the lens.

B: Draw a line representing the **principal axis** and mark the positions of the principal foci (shown as F on the diagram).

C: Draw **three** rays from the top of the object:

   1 – a ray parallel to the principal axis that is refracted by the lens so that it passes through the principal focus on the far side of the lens

   2 – a ray that passes through the principal focus on the same side of the lens as the object, to the lens and is then is refracted parallel to the principal axis

   3 – a ray that passes through the centre of the lens without being refracted.

D: Draw in the real image that is formed where the three rays meet.

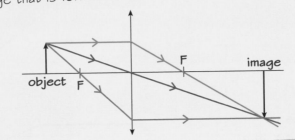

Worked example

Describe the image formed by the lens in the ray diagram above.
(3 marks)

The image is real, inverted and magnified.

Now try this

1 A lens in a telescope produces an image of a star on a screen in a digital camera. State whether the image is real or virtual, and whether it is bigger or smaller than the object.
(2 marks)

Real images in lenses

A converging lens produces a real image when the object is placed more than one focal length away from the lens. The size and position of the image depend on the distance of the object from the lens.

Worked example

target B-A*

AQA SKILL
Draw
Page 79

1 A student wants to get an image of a candle flame on a screen. He places a convex lens 10 cm from the candle. The lens has a focal length of 8 cm.

(a) Complete the ray diagram to find where the image is formed. *(3 marks)*

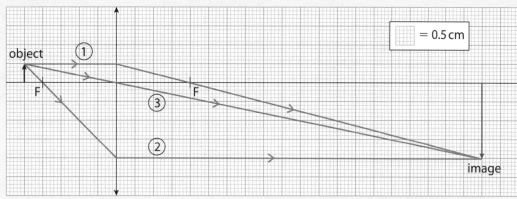

(b) Calculate the magnification of the image.

image height = 2 cm

object height = 0.5 cm

$$\text{magnification} = \frac{2\,\text{cm}}{0.5\,\text{cm}} = 4$$

$$\text{magnification} = \frac{\text{image height}}{\text{object height}}$$

EXAM ALERT!

Always use a pencil and ruler when drawing ray diagrams – and don't forget the arrows to show the direction of the rays.

Students have struggled with exam questions similar to this – **be prepared!**

Now try this

target D-C

target B-A*

1 A convex lens forms an image 6 cm high. The object is 8 cm high. What is the magnification? *(2 marks)*

2 Complete the ray diagram below to find the position and nature of the image. *(4 marks)*

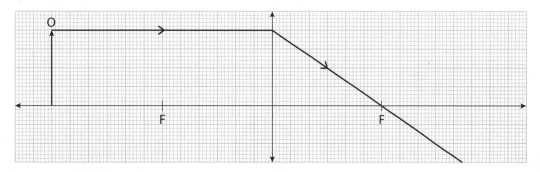

Virtual images in lenses

Converging lenses form virtual images when the object is closer than one focal length. Diverging lenses always form virtual images.

Magnifying glass

A converging lens acts as a MAGNIFYING GLASS when the object is between the principal focus and the lens.

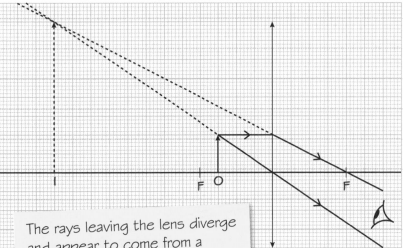

The rays leaving the lens diverge and appear to come from a point on the same side of the lens as the object. This is a **virtual** image, which is upright and magnified.

Concave lenses

CONCAVE lenses are thinner in the middle than at the edges. They are also called DIVERGING lenses because they make a beam of light spread out.

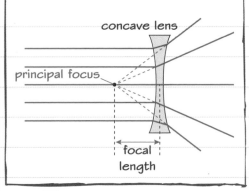

A concave lens can be shown as)(.

 Worked example   B-A*

An object is 8 cm from a concave lens with a focal length of 5 cm. Construct a ray diagram showing the formation of the image and state the nature of the image.

The image is virtual, upright and smaller than the object.

The image formed by a diverging lens is *always* virtual, upright and diminished.

Use dashed lines for showing rays extended to cross at virtual images.

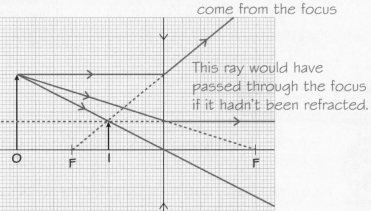

This ray appears to have come from the focus

This ray would have passed through the focus if it hadn't been refracted.

 Now try this

D-C   1 Explain why a concave lens can never be used to project an image onto a screen.
(2 marks)

B-A*   2 A magnifying glass with a focal length of 8 cm is used to give a clear image of an object 4 cm from the lens. Construct a ray diagram to find the magnification of the image. *(4 marks)*

The eye

The eye produces images of objects at various distances.

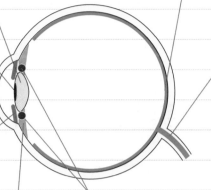

LENS – can change shape to help focus light to form image on the retina.

RETINA – made up of cells sensitive to light. Sends impulses to the brain.

PUPIL – hole in the centre of the iris through which light passes.

CORNEA – transparent so lets light enter the eye and helps to focus it onto the retina.

OPTIC NERVE

IRIS – adjusts size of the pupil to control the amount of light entering the eye

CILIARY MUSCLES – contract and relax to change the shape of the lens.

SUSPENSORY LIGAMENTS – hold the ciliary muscles and lens in place.

The cornea and the lens together bend the light to form a real image on the retina.

Worked example

target D-C

The diagrams show the eye in two different lighting conditions. Explain the differences shown in the diagrams.

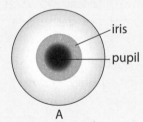

iris

pupil

A

B

Diagram A shows the eye in dim light. The pupil is large to allow plenty of light to get into the eye.

Diagram B shows the eye in bright light. The pupil is small. This prevents too much light getting into the eye, which might damage the retina.

Now try this

target D-C

1 Which **two** of the following parts of the eye are involved in bending light to form an image?   *(2 marks)*

| cornea | lens | pupil | retina |

target D-C

2 Which part or parts of the eye:
(a) control the amount of light entering the eye?   *(1 mark)*
(b) help the eye to focus on objects at different distances?   *(1 mark)*
(c) convert light into electrical impulses?   *(1 mark)*

Range of vision

The ciliary muscles change the shape of the lens to allow the eye to focus on objects at various distances.

Viewing distant objects

Light from distant objects does not need to be bent very much to form an image on the retina. The lens is thin.

The furthest distance at which the eye can focus light is called the FAR-POINT and is an infinite distance away from the eye.

The ring of ciliary muscle relaxes and the lens is pulled into a thin shape.

Viewing close objects

Light from close objects needs to converge more, so the lens needs to be fatter.

The ring of ciliary muscle contracts and the lens becomes rounder.

The closest distance from the eye at which a sharp image can be formed is called the NEAR-POINT and is about 25 cm from the eye.

Worked example   D-C

The diagram shows how a camera works. Describe the similarities and differences between the eye and a camera.   *(6 marks)*

The aperture (opening) can be changed to allow different amounts of light into the camera.

The lens focuses light onto the film.

The lens can be moved to adjust the focus for objects at different distances.

Chemical changes occur in the film when light hits it. These changes are made permanent when the film is developed.

A shutter stops light getting to the film. The shutter is opened when you take a picture. The length of time the shutter is open can be adjusted.

Similarities:
- The size of the pupil is adjusted by the iris to control the amount of light entering the eye. A camera has a similar arrangement.
- The eye and the camera both have a convex lens for focusing light to form a real image.

Differences:
- In the eye, the shape of the lens changes to focus on objects at different distance. In the camera the position of the lens is changed.
- Cameras also have a shutter which allows light to enter for a limited time. The eye does not need this.
- The camera records the image on a film or CCD (charged couple device) while in the eye the image formed on the retina is converted into nerve signals sent to the brain.

Now try this

B-A

1 Describe what happens in the eye when it adjusts from looking at a distant object to an object about 25 cm from the eye.   *(3 marks)*

Correction of sight problems

Lenses are used to correct problems in forming sharp images in the eye.

Long sight

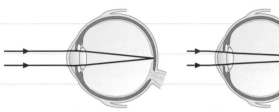

| normal eye | long sight |

Long sight occurs when the eyeball is shorter than normal or the lens is flatter than normal, so that light from close objects cannot be focused on the retina in a sharp image.

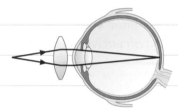

Convex lenses correct long sight.

Long sight is corrected using converging lenses. These bend the light more so that the rays meet to form an image on the retina.

Short sight

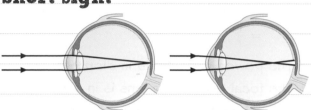

| normal eye | short sight |

Short sight occurs when the eyeball is longer than normal or the lens is more curved than normal, so that light from distant objects forms a blurred image on the retina.

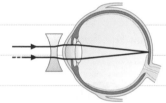

Concave lenses correct short sight.

Short sight is corrected using diverging lenses. These spread the light before it enters the eye and is converged by the cornea and lens onto the retina.

Worked example   target B–A*

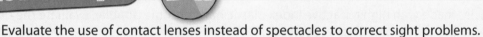

AQA SKILL Evaluate Page 73

Evaluate the use of contact lenses instead of spectacles to correct sight problems.   *(3 marks)*

An advantage of using contact lenses is that they are less likely to be knocked or damaged when the wearer is taking part in sport. A disadvantage is that contact lenses must be kept clean to prevent damage to the cornea. Overall contact lenses are a good idea if you are taking part in lots of sport and are prepared to keep them clean.

> Contact lenses rest on the cornea. Particles of dust or microbes on the lens could harm the cells that make up the cornea.

Now try this

 target D–C

1 A person can see close objects clearly but cannot focus on distant objects.
  (a) Is the person long sighted or short sighted?   *(1 mark)*
  (b) Describe **one** possible cause of this condition.   *(1 mark)*
  (c) Explain which type of lens should be used in spectacles for this person.   *(2 marks)*

 target B–A*

2 As people grow older they often have difficulty focusing their eyes on print in books. Explain which kind of lens is used in reading glasses.   *(2 marks)*

Power of a lens

The power of a lens is its ability to bend light.

The power of a lens is related to its focal length.

$$\text{power} = \frac{1}{\text{focal length}}$$

$$= \frac{1}{f}$$

- The power of a lens is in dioptres, D
- The focal length of a lens is in metres, m

The focal length and power of a convex (converging) lens are positive. The focal length of a concave (diverging) lens is negative, so the power is also negative.

A beam of light spreads out after passing through a concave lens so it only *appears* to have come from the focal point.

The focal length of a lens depends on:

- the refractive index of the material (the higher the refractive index the shorter the focal length)
- the curvature of the two sides of the lens (the more curved the lens, the shorter the focal length).

The term 'power' used here is not related to power measured in watts.

Refractive index   HIGHER

The higher the refractive index of a material, the more the light is bent. To achieve a certain focal length, a lens with a higher refractive index will not have to be as curved and so can be flatter and slimmer than a lens made with a lower refractive index material.

Worked example

1 (a) A converging lens has a focal length of 10 cm. Calculate the power of the lens. *(2 marks)*

P = 1/0.1 = 10 D

The focal length must be in metres. 10 cm is 0.1 m.

(b) The focal length of a concave lens is −0.25 m. Calculate the power of the lens. *(2 marks)*

P = 1/−0.25 = −4 D

Don't forget to include the minus sign.

2 Spectacles made with high refractive index plastic have become popular. Evaluate the use of high refractive index materials in spectacles. *(3 marks)*

Using high refractive index plastic means that the focal length needed can be achieved with less curve on the lens. This means that the lens can be thinner and the spectacles can be lighter and look better. However, the plastic is expensive and scratches more easily than glass. Most people would buy the high refractive index lenses if they can afford them.

Now try this

1 Choose which of the lenses shown on the right will have the shortest focal length and explain your answer. *(2 marks)*

2 Lens X has a power of 5 dioptres and lens Y has a power of 5.4 dioptres. The two lenses are the same shape and size. Explain which lens is made of the material with the higher refractive index. *(3 marks)*

A : refractive index 1.5

B : refractive index 1.5

C : refractive index 1.4

Total internal reflection

Reflection and refraction happen when light hits the boundary between media.

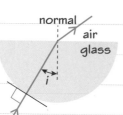

increasing the angle i →

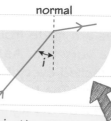

A ray of light enters the block of glass along the normal so it is not refracted at the first face. It is refracted by the straight side of the glass and emerges into the air.

The ray in the glass hits the boundary at a larger angle and is refracted at a bigger angle.

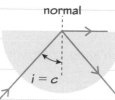

angle i = the critical angle c

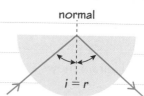

total internal reflection
angle i > critical angle c

The ray hits the boundary between glass and air at the **critical angle**. The angle of refraction in the air is 90° so some of the light travels along the boundary and some is reflected.

When i > c, the light is totally internally reflected. Remember that **total internal reflection** only occurs when light is leaving a material such as glass or air.

Worked example

target **B-A***   HIGHER

In an optical fibre the critical angle at which total internal reflection takes place is 39°. Calculate the refractive index of the glass.

Refractive index = $\dfrac{1}{\sin 39} = \dfrac{1}{0.629} = 1.59$

refractive index = $\dfrac{1}{\sin c}$

c is the critical angle.

Make sure you can use your calculator to find the 'sin' of an angle.

Now try this

target **G-E**

1 Complete the sentence below using words from the box.

| air   glass   the normal |

Total internal reflection can occur when light goes from ………….. to ………….. but not when it is going from ………….. to …………..
(2 marks)

target **B-A***

2 Calculate the critical angle for glass that has a refractive index of 1.5. (3 marks)

HIGHER

Other uses of light

Optical fibres and lasers make use of properties of light.

Optical fibres

Optical fibres are made from a special glass. They can bend because the fibres are very thin. As the ray of light travels down the glass fibre it is reflected every time it hits the edge of the fibre, even if the fibre is bent. The law of reflection is obeyed at each contact with the interface. The light does not escape from the fibre.

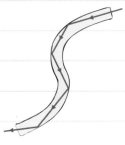

Worked example   target **D-C**

1 (a) State how optical fibres are used in communications. *(1 mark)*

Optical fibres are used to carry telephone and TV signals.

(b) The diagram shows a picture of an endoscope. Explain how an endoscope works. *(3 marks)*

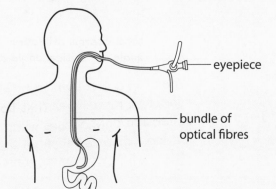
— eyepiece
— bundle of optical fibres

The endoscope is an instrument used to look at the inside of the body. It contains a bundle of optical fibres that carry light into and back out of the body to form an image.

Lasers

A laser produces a narrow, intense beam of light of a single wavelength. The laser beam can concentrate a lot of energy into a small area. This means it can burn or cut through materials very precisely. It can also be used for CAUTERISING (sealing wounds by heating them).

The surgery is expensive and there is always a risk of infection.

Worked example   target **D-C**

2 Explain how lasers are used to correct defects in sight. *(2 marks)*

A very narrow laser beam can make very fine cuts in the cornea of the eye in order to change its shape and correct long or short sight.

Now try this

target **D-C**

1 Describe the route that light takes when a doctor uses an endoscope to look at the inside of a patient's stomach. *(3 marks)*

target **B-A***

2 Evaluate the use of laser surgery for correcting long or short sight. *(3 marks)*

Physics six mark question 1

There will be one six mark question on your exam paper, which will be marked for *quality of written communication* as well as scientific knowledge. This means that you need to apply your scientific knowledge, present your answer in a logical and organised way and make sure that your spelling, grammar and punctuation are as good as you can make them.

Worked example

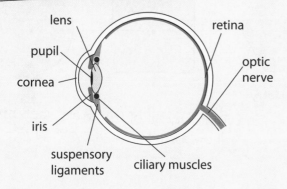

lens
retina
pupil
optic nerve
cornea
iris
suspensory ligaments
ciliary muscles

The human eye is very well adapted to forming images of objects both distant from and close to the eye.

Explain how the parts of the human eye are involved in forming sharp images of distant and near objects.   *(6 marks)*

Light enters the eye through the cornea which acts as a convex lens. The rays converge and pass through the pupil. The iris adjusts the size of the pupil to control the amount of light entering the eye.

The lens converges the light to form a sharp image on the retina.

When the object being viewed is distant the ciliary muscles are relaxed and the lens is pulled thin by the suspensory ligaments.

When the object is close to the eye the rays must be converged more. The ciliary muscles contract, squeezing the lens into a rounder, more curved shape which bends the light rays more. The near point is the distance to the eye from the closest point where a sharp image can be obtained. It is normally about 25 cm.

Explain

Look at the question carefully – it asks how the different parts of the eye work and about focusing on near and distant objects. The command word is EXPLAIN, which means you must give reasons for what happens in the eye linked together logically.

This is a good answer because it explains how an image is formed on the retina. The diagram will help you but you need to know the function of all the parts of the eye.

The second part of the answer then explains how the eye adjusts to produce sharp images of distant and near objects.

Now try this

Compare the properties of X-rays and ultrasound and their uses in medicine.
(6 marks)

Compare means that you must give similarities and differences between the two waves.

Centre of mass

All objects have a centre of mass no matter what shape they are.

The CENTRE OF MASS of an object is the point at which the whole mass of the object may be thought to be concentrated.

When you lift an object the force of gravity appears to act through the centre of mass.

centre of mass

The centre of mass of an object may be outside the object, such as this boomerang.

Regular objects

For a regular shaped object the centre of mass lies on the axes of symmetry.

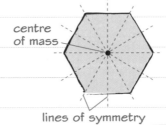

centre of mass

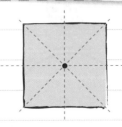

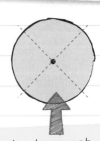

lines of symmetry

The diagram of the circle does not show all the axes of symmetry.

1 Describe how you can find the centre of mass of a thin, irregular shaped sheet of material such as is shown on the right. *(3 marks)*

1. Fix a thread to the object and let it hang freely from a hook.
2. Hang a plumb line (a weight on a string) from the same hook.
3. Mark the line of the plumb line on the object.
4. Move the thread to two more different points on the object and repeat steps 1 to 3.
5. The centre of mass is the point where the three lines cross.

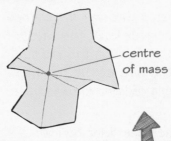

centre of mass

If you are asked to describe how to carry out a practical procedure like this, it helps to number the different steps.

When an object is suspended freely it comes to rest with its centre of mass on a line vertically below the point of suspension.

1 Mark the position of the centre of mass of the shapes shown below. *(4 marks)*

2 A shopkeeper has the letters of his name made in wood. He wants to fix each letter with a single nail through the centre of mass. The shopkeeper's name is CHILDS. Explain why some letters cannot be fixed with a single nail. *(2 marks)*

The pendulum

A simple PENDULUM consists of an object, called a bob, hanging by a cord from a fixed point.

The PERIOD of a simple pendulum is the time taken for a complete swing.

The FREQUENCY of a pendulum is the number of swings completed in one second.

Period = 1/frequency

$T = 1/f$

- T is the periodic time, in seconds, s
- f is the frequency, in Hertz, Hz

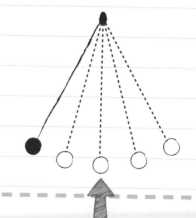

Length of the pendulum

The period of a simple pendulum increases with its length. The length of a pendulum is the distance from the pivot to its centre of mass which is usually in the middle of the mass hanging from the cord.

A complete swing is from the starting point out to the furthest extent of the swing and back again.

Worked example **D-C**

1 The Pirate Boat is a popular attraction at many amusement parks. When it is swinging freely this Pirate Boat passes the bottom point of its swing 24 times in 1 minute.

(a) Calculate the frequency of the ride in Hz. *(2 marks)*

The boat passes through the bottom point twice on each complete swing. The frequency is 24/2 = 12 swings per minute (or per 60 seconds).

Frequency = 12/60 = 0.2 Hz.

(b) Calculate the period of the ride. *(2 marks)*

T = 1/0.2 Hz = 5 s

2 A student is trying to make a pendulum that has a period of exactly 1 s. The pendulum has a period of 1.2 s. Explain what the student needs to do to change the period of the pendulum. *(2 marks)*

The student should shorten the length of the pendulum. This is because the period decreases when the distance between the pendulum and the mass decreases.

Now try this

 D-C

1 A child on a swing in a playground is pulled back and then released. The child completes 30 complete swings in a minute. Calculate:

(a) the frequency in Hz *(2 marks)*

(b) the period of the swing in seconds. *(2 marks)*

B-A* 2 An adult takes the place of the child on a swing. Explain what happens to the frequency or period of the swing. *(3 marks)*

Turning effect and levers

If you hold a metre rule near one end it feels as if the object is trying to turn. That is due to the turning effect or MOMENT of the object's weight.

Moments

The size of the moment, or turning effect, of a force is given by the equation:

Moment = force × perpendicular distance to pivot

$M = F \times d$

- The moment, M, is in newton-metres, Nm.
- Force, F, is in newtons, N.
- The distance, d, is in metres, m. The distance must be measured from the pivot perpendicular (at right angles) to the line of action of the force.

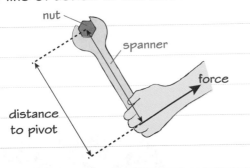

nut

spanner

force

distance to pivot

Levers

A LEVER is a machine that uses moments to move objects.

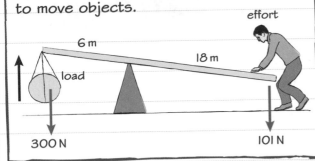

effort

6 m

18 m

load

300 N

101 N

The effort needed is much less than the load. Levers are called force multipliers.

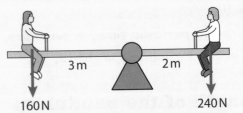

Worked example   D-C

1 Two children are sitting on a seesaw.

(a) Calculate the anticlockwise moment. *(2 marks)*

3 m   2 m

160 N   240 N

anticlockwise moment (on the left)
= 160 N × 3 m = 480 Nm

(b) Calculate the clockwise moment. *(2 marks)*

Clockwise moment (on the right)
= 240 N × 2 m = 480 Nm

(c) Explain why the seesaw doesn't move. *(2 marks)*

The anticlockwise and clockwise moments about the pivot are equal so the seesaw is balanced and does not move.

2 Look at the diagram of a lever on the left. Explain why the load in the diagram is being lifted. *(3 marks)*

For the load to rise the clockwise moment of the effort must be bigger than the anticlockwise moment of the load.

- anticlockwise moment
  = 300 N × 6 m = 1800 Nm
- clockwise moment
  = 101 N × 18 m = 1818 Nm

So the load will be lifted.

Now try this

 D-C

1 Look at the diagram of the spanner at the top of the page.

(a) The force is 8 N and the distance shown labelled in the diagram is 0.12 m. Calculate the moment being used to turn the spanner and state its direction. *(3 marks)*

(b) The nut is not moving. State the moment of the force acting on the nut. *(1 mark)*

2 Look at the diagram of the lever above. Suggest **two** changes you could make to reduce the force needed to lift the 300 N load. *(2 marks)*

B-A*

Moments and balance

HIGHER   This whole page is Higher material.

When an object is balanced, the clockwise moment is balanced by the anticlockwise moment.

When an object is balanced it is possible to use moments to work out the size of a force, or the distance from the force to the pivot.

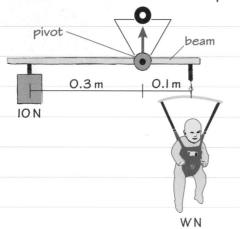

pivot

beam

0.3 m   0.1 m

10 N

W N

The diagram shows a common arrangement for weighing fruit and vegetables, as well as babies. The beam is balanced so:

anticlockwise moment = clockwise moment

$10\,N \times 0.3\,m = W\,N \times 0.1\,m$

$W = 3\,Nm/0.1\,m = 30\,N$

 The weight of the bar has been ignored in this calculation.

Worked example   target B-A*

1 A light shelf is held by a fixing on the wall. The fixing will break if the force on it is greater than 6 N. Calculate how far from the wall an ornament with a weight of 2 N can be placed without the shelf collapsing.   *(3 marks)*

The shelf is balanced when the ornament is at its maximum distance from the pivot so:

anticlockwise moments = clockwise moments.

$2\,N \times d = 6\,N \times 5\,cm$

$d = 30/2 = 15\,cm$ from pivot or 20 cm from the wall.

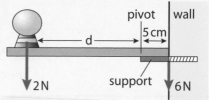

pivot | wall
5 cm

d

2 N   support   6 N

 The units of length can be m or cm but must be the same for all the moments in the calculation.

2 A ladder is resting against a wall. The weight of the ladder is 20 N. Calculate the force that the wall applies to the top of the ladder.

The ladder is pivoting about its base.

Moment caused by the weight of the ladder (anticlockwise moment) = $20\,N \times 0.5\,m = 10\,Nm$

The ladder is not moving, so the clockwise moment (caused by force F) = 10 Nm.

$F = 10\,Nm/1.8\,m = 5.56\,N$

Remember that when you are calculating moments, the distance must be perpendicular to the force.

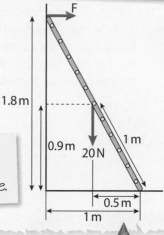

F

1.8 m

0.9 m   20 N   1 m

0.5 m

1 m

Now try this   target B-A*

1 A uniform seesaw 3 m long is pivoted at its mid-point. A child weighing 160 N sits at one end. Where must a child with a weight of 200 N sit to balance the seesaw?   *(3 marks)*

EXAM ALERT!

Remember to show every stage in your calculation.

Students have struggled with exam questions similar to this – **be prepared!**

Stability

A STABLE object is one that is not likely to topple over. The stability of an object depends on the position of its centre of mass and the width of its base.

A Formula 1 racing car is much more difficult to tip over than an ordinary car because its centre of mass is much lower. Its wheels are wide apart so it has a wider base than an ordinary car.

The lower the centre of mass, and the wider the base, the greater the stability of an object. A stable object has to be tilted a long way to make it fall over.

Line of action HIGHER

An object is stable if the line of action of its weight passes through the base. If the line of action of the weight is outside the base there will be a moment around the edge of the base and the object will topple over.

There is a clockwise moment caused by the weight, which stops the bus toppling over.

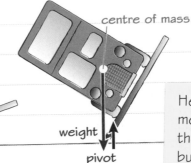

centre of mass

centre of mass

weight

pivot

weight

pivot

Here the anticlockwise moment caused by the weight makes the bus topple over.

Worked example G-E B-A* HIGHER

AQA SKILL
Suggest
Page 79

1 A café owner serves drinks in glasses like the one shown in the diagram. Many customers accidently knock over their drinks. Suggest **two** ways the design could be improved. *(2 marks)*

The glass should have a wider base and the base should be heavier so that the centre of mass is lower.

2 Look at the diagram of the glasses shown on the left. Explain why this design means that the glass is not very stable. *(2 marks)*

When the glass is full, the centre of mass is high so even a small angle of tilt will bring the line of action of the weight outside the narrow base. There will be a moment about the point of contact that will make the glass topple over.

Now try this

D-C

B-A*

HIGHER

1 Look at the pictures of vases of flowers.

Explain which of the vases is least likely to topple over. *(3 marks)*

2 Explain why it is difficult to balance on a stationary bicycle without toppling over. *(2 marks)*

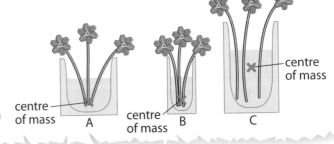

centre of mass A

centre of mass B

centre of mass C

Hydraulics

HYDRAULIC systems use a liquid to transfer forces from one place to another. It is almost impossible to compress a liquid into a smaller volume. When PRESSURE is exerted on a liquid the same pressure is transmitted equally in all directions.

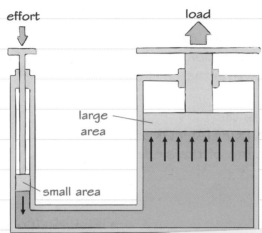

effort

load

large area

small area

Pressure is the force acting on unit area.

A hydraulic system works as a force multiplier.

Worked example   target D-C

In the diagram above the effort force of 5 N acts on an area of 0.01 m².

(a) Calculate the pressure in the liquid under the effort piston. **(2 marks)**

pressure = 5 N/0.01 m² = 500 Pa

(b) What is the pressure on the load piston? **(1 mark)**

500 Pa

The pressure in a hydraulic system is given by the equation

$$\text{pressure} = \frac{\text{force}}{\text{area}}$$

$$P = \frac{F}{a}$$

- Pressure, P, is in pascals, Pa.
- Force, F, is in newtons, N.
- Area, a, is in metres squared, m².

EXAM ALERT!

In hydraulics, remember that the same **pressure** is transmitted through the liquid, not the same **force** – the force may well be larger.

Students have struggled with exam questions similar to this – **be prepared!**

Now try this

target D-C

1 In a garage a hydraulic jack is used to lift a car that has a weight of 8000 N. The area of the piston holding the car is 0.04 m².

(a) Calculate the pressure in the hydraulic liquid. *(2 marks)*

(b) Explain how one man can use a force of just 50 N to raise the car. *(2 marks)*

target B-A*

2 When the foot-brake in a car is pushed down the force is transmitted by a hydraulic system to the brakes. The cross-sectional area of the piston connected to the foot-brake is 0.0001 m² and that of the piston connected to the brakes is 0.002 m². Calculate the force that must be applied to the foot-brake to produce a braking force of 160 N. *(3 marks)*

Circular motion

A force is needed to make an object move in a circle.

Velocity, acceleration and force are vectors – they have direction as well as size.

When an object is moving in a circle its direction is constantly changing. This means that it is accelerating. The acceleration of an object moving in a circle is towards the centre of the circle.

The resultant force that causes the acceleration is called the CENTRIPETAL FORCE. It always acts towards the centre of the circle.

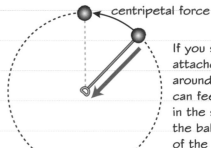
centripetal force

If you swing a ball attached to a string around your head you can feel the tension in the string pulling the ball to the centre of the circle.

EXAM ALERT!

Remember **centripetal** is not a type of force like friction or tension – it is the name given to the resultant force that causes an object to go in a circular path.

Students have struggled with this topic in recent exams – **be prepared!**

You should be able to name the type of force that produces the centripetal force keeping something moving in a circle. For example, gravity keeps the Earth moving around the sun, friction between the tyres and the road makes a car turn a corner.

Worked example   D-C

In the Olympic sport of hammer throwing the mass is attached to a cord 1.2 m long. The hammer is swung around in a circle before it is released.

(a) Explain what happens to the hammer when the thrower releases it.   *(2 marks)*

The hammer flies off in a straight line because the centripetal force is no longer acting.

(b) State **three** factors that can increase the centripetal force acting on the hammer.   *(3 marks)*

Increasing the mass of the hammer. Increasing the speed the hammer rotates. Making the radius of the circle smaller by making the cord shorter.

Now try this

D-C

1 The photo and diagram show a fairground ride.

  (a) Name the resultant force that keeps the swings moving in a circle.   *(1 mark)*

  (b) Mark on the diagram the direction of the resultant force.   *(1 mark)*

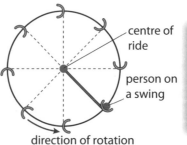
centre of ride

person on a swing

direction of rotation

  (c) Explain why the force on an adult during the ride will be different than the force on a child.   *(2 marks)*

B-A* **2** Explain why cyclists have to take corners more slowly when the road is icy.   *(3 marks)*

Physics six mark question 2

There will be one six mark question on your exam paper, which will be marked for *quality of written communication* as well as scientific knowledge. This means that you need to apply your scientific knowledge, present your answer in a logical and organised way and make sure that your spelling, grammar and punctuation are as good as you can make them.

Worked example

A sauce manufacturer is running trials of a new sauce. They have provided samples in two bottles shown in the diagram. The bottles have the same volume. Tests show that bottle A tends to fall over more often than bottle B, but shopkeepers said they could get more of bottle A on a shelf.

Evaluate the design of the two sauce bottles.

(6 marks)

Bottle A has the smaller cross-sectional area so takes up the least room on a shelf. This means that more can be packed together.

The centre of mass of the bottles is the point at which all their mass seems to be concentrated. The centre of mass is where the lines of symmetry cross. The centre of mass of bottle A will be higher than for bottle B. Bottle A is more unstable than bottle B because of its narrow base and high centre of mass.

The advantage of bottle B is that it is more stable, but the disadvantage is that you can fit fewer on the shelf.

To avoid damage to the bottles it would be better to use bottle B.

Higher-tier students may note that the bottles fall over when the line of action of the centre of mass falls outside the base. This happens for a smaller angle of tilt with bottle A than bottle B.

Evaluate

The instruction EVALUATE means you should consider the advantages and disadvantages of the two designs, giving reasons, and state your conclusion.

Use the evidence given in the question as part of your answer. Bottle A has an advantage as well as a disadvantage.

Now try this

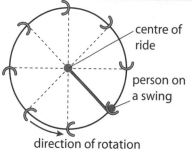

- centre of ride
- person on a swing
- direction of rotation

The picture shows a fairground ride. People sit in seats hanging by chains that are pulled in a circle. The ride is an old one and has to be inspected regularly to check that it is safe and that a chain holding one of the seats will not break. Discuss the factors that would increase the centripetal force on the ride and make it more likely for the chains to break. *(6 marks)*

Electromagnets

Electromagnets are used in a variety of applications, such as cranes, relays and circuit breakers.

When an electric current travels along a wire it produces a magnetic field. If the wire is wound into a coil, the shape of the magnetic field is similar to that of a bar magnet.

The wire is usually wound around a bar made of soft iron.

The force can be increased by increasing the current and the number of turns of wire.

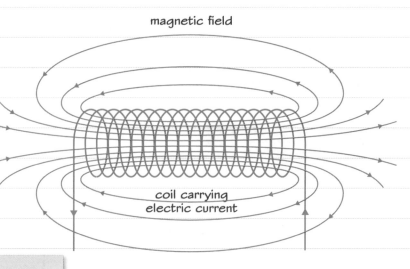

magnetic field

coil carrying electric current

EXAM ALERT!

Remember the factors that affect the strength of an electromagnet.

Students have struggled with this topic in recent exams – **be prepared!**

Worked example

B–A*

AQA SKILL
Suggest
Page 29

The diagram shows a circuit breaker, which is used to stop large currents flowing in a circuit. The switch is normally held closed by a spring (not shown here). Suggest how the circuit breaker works.   *(4 marks)*

When the current in the circuit is normal, the spring is strong enough to hold the switch closed. If the current increases, the force from the electromagnet increases enough to pull the switch open. The circuit is broken and the current cannot flow.

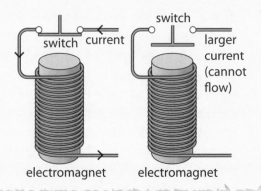

switch current

switch

larger current (cannot flow)

electromagnet   electromagnet

Now try this

target
D–C

1 Explain why an electromagnet is more useful than a permanent magnet for moving steel objects around in a scrap yard.   *(3 marks)*

target
B–A*

2 The diagram on the right shows a relay. Relays are often used to control circuits carrying large currents. Explain how relays are used in this way.   *(2 marks)*

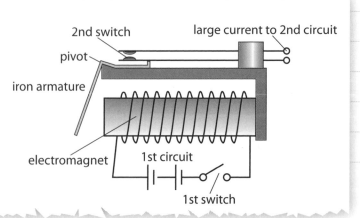

2nd switch   large current to 2nd circuit

pivot

iron armature

electromagnet   1st circuit

1st switch

The motor effect

An electric current flowing through a wire has a magnetic field. If you put the wire into the field of another magnet the two fields affect each other and the wire experiences a force. This is called the MOTOR EFFECT.

The maximum force on the wire occurs when the current is at right angles to the lines of the magnetic field. We can work out the direction of the force using Fleming's Left Hand Rule.

current

field

N   S

movement

First finger
Field

seCond finger
Current

thuMb
Movement

This rule uses 'conventional current' which flows from the + to the − of a cell. This is the opposite direction to the flow of electrons.

- There is no force if the current is parallel to the field lines.
- If the direction of either the current or the magnetic field is reversed the direction of the force is reversed.
- The size of the force can be increased by increasing the strength of the magnetic field, or increasing the size of the current.

Worked example   target B-A*

1 An electric motor uses the motor effect to make a motor go round. Use the diagram to help you answer the questions.

(a) Explain why the left-hand side of the coil moves upwards.
(2 marks)

The left-hand side of the coil is pushed upwards by the motor effect. The magnetic field runs from left to right, the current is running out of the page and we can use Fleming's Left Hand rule to work out that the direction of the force is upwards.

(b) Explain why the coil rotates.   *(2 marks)*

The motor effect means that the left-hand part of the coil moves up and the right-hand side moves down. This makes the coil spin. The split ring swaps the connections to the battery over every half turn, so the coil continues to spin in the same direction.

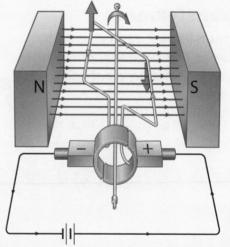

Now try this

1 Look at the diagram of a current along a wire in a magnetic field. State the direction the wire will move:

(a) in the diagram

(b) if the poles of the magnet are reversed

(c) if the current in the wire is reversed.   *(3 marks)*

The point shows the current is towards you.

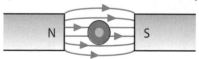

N   S

2 The electric motors in a railway locomotive and a food mixer work on the same principles. Give **one** similarity and **one** difference between the two electric motors.   *(2 marks)*

71

Electromagnetic induction

A potential difference is induced in a wire if the wire is moved through a magnetic field.

When an electrical conductor moves in a magnetic field it 'cuts' across the lines of force. This induces a potential difference in the wire and an electric current flows through the circuit.

> A current is also induced if the magnetic field moves and the wire, or a coil of wire, stays motionless.

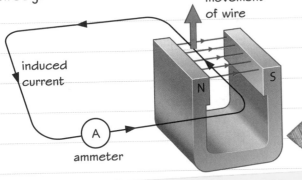

induced current

ammeter

movement of wire

Induce means to create or make something.

EXAM ALERT!

Remember that a conductor must be in a **changing** magnetic field for a potential difference to be induced across it.

Students have struggled with this topic in recent exams – **be prepared!**

Transformers

A transformer can be used to change the potential difference of an alternating electricity supply.

> Remember that the primary coil is the one connected to the electricity supply.

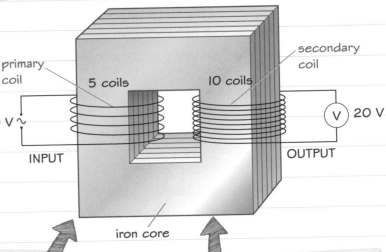

primary coil

5 coils   10 coils

secondary coil

10 V ~

INPUT

V 20 V

OUTPUT

iron core

When an alternating current flows through the primary coil it produces a changing magnetic field in the iron core. This magnetic field induces an alternating potential difference across the secondary coil.

The core must be made of a magnetic material so that it can channel the magnetic lines of force from the primary coil to the secondary coil.

Now try this

target D–C

1 If the primary coil of a transformer is connected to a battery the reading on the voltmeter connected across the secondary coil is 0 V. Explain this observation. *(3 marks)*

target B–A*

2 No part of a transformer moves. Explain how a potential difference is induced in the secondary coil. *(2 marks)*

Step-up and step-down transformers

Transformers are used to increase or decrease the potential difference of an alternating electricity supply.

In a STEP-UP transformer the potential difference across the secondary coil is greater than the potential difference across the primary coil.

In a STEP-DOWN transformer the potential difference across the secondary coil is less than the potential difference across the primary coil.

The change in potential difference between the primary and secondary coils is determined by the ratio of the number of turns of wire on the primary and secondary coils.

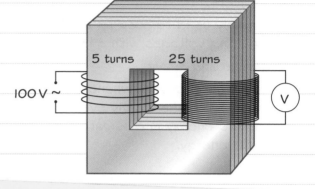
5 turns   25 turns
100 V ~   (V)

The relationship between the potential difference in the primary and secondary coils is given by the equation

$$\frac{V_p}{V_s} = \frac{n_p}{n_s}$$

- V_p is the potential difference across the primary coil in volts, V.
- V_s is the potential difference across the secondary coil in volts, V.
- n_p is the number of turns in the primary coil.
- n_s is the number of turns in the secondary coil.

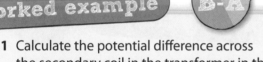

Worked example   B-A*

1 Calculate the potential difference across the secondary coil in the transformer in the diagram above.   *(3 marks)*

$$V_s = V_p \times \frac{n_s}{n_p} = 100 \text{ V} \times \frac{25}{5}$$

$$= 500 \text{ V}$$

It is assumed that transformers are 100% efficient. This means that the input power of the primary coil and the power output of the secondary coil (called the LOAD) are equal.

For a transformer:

$$V_p \times I_p = V_s \times I_s$$

- V_p and V_s are the potential difference across the primary and secondary coils in volts, V.
- I_p and I_s are the current in the primary coil and secondary coils in amps, A.

Remember, for electrical power: $P = I \times V$.

Worked example   B-A*

2 In the transformer shown above, the current in the primary coil is 1.5 A. Calculate the current in the secondary coil assuming the transformer is 100% efficient.   *(3 marks)*

$$I_s = V_p \times \frac{I_p}{V_s}$$

$$= 100 \text{ V} \times \frac{1.5 \text{ A}}{500 \text{ V}}$$

$$= 0.3 \text{ A}$$

Now try this

D-C

1 A transformer has 50 turns of wire on the primary coil and 200 on the secondary. The potential difference across the primary coil is 12 V.
Explain why the output potential difference is 48 V.   *(2 marks)*

B-A*

2 A transformer with an input potential difference of 30 kV has a load on the secondary coil of 3 kW. What is the current in the primary coil? State the assumption that you make.   *(3 marks)*

Switch mode transformers

Switch mode transformers are used to supply power to electronic appliances.

Traditional transformers

Traditional transformers have a primary and secondary coil wound on an iron core. These transformers:

- are mainly used to step up or down mains alternating voltages with a frequency of 50 Hz
- are large and heavy
- use electrical energy whenever they are connected to the supply even if there is no load. The energy is dissipated as heat energy.

A transformer used to convert 230 V to 110 V for use with power tools. This transformer dissipates energy whenever it is plugged in and switched on, even if the power tool is not being used.

Switch mode transformers

SWITCH MODE TRANSFORMERS use electronic components to convert alternating current to direct current. The direct current is then turned on and off at a very high frequency (50–200 kHz). The output is a current at a lower potential difference than the mains.

A switch mode transformer.

1 The chargers for portable phones are usually left plugged into the mains power. State why switch mode transformers ensure that energy is not wasted when this is done.

(1 mark)

Switch mode transformers use very little power when the battery in the phone is fully charged and there is no load.

2 Early laptop computers used traditional transformers to charge their batteries. Modern laptops use switch mode transformers to do the same task. Explain why the change in transformer has been made. *(2 marks)*

Switch mode transformers are much lighter and smaller than traditional transformers so are more convenient to carry. Also, they use very little power when the laptop batteries are fully charged, so less energy is wasted.

Now try this

1 Give **three** differences between traditional transformers and switch mode transformers.
(3 marks)

2 The electronics in appliances such as televisions need a different potential difference to mains electricity. Old-style televisions on 'stand-by' used up to 25% of the power used when on. Explain why this doesn't happen with modern televisions. *(2 marks)*

Physics six mark question 3

There will be one six mark question on your exam paper, which will be marked for *quality of written communication* as well as scientific knowledge. This means that you need to apply your scientific knowledge, present your answer in a logical and organised way and make sure that your spelling, grammar and punctuation are as good as you can make them.

Worked example

The diagram shows the circuit and construction of an electric bell. When the switch is closed the striker hits the gong repeatedly, making the ringing sound.

Suggest how pressing the switch makes the bell operate.   *(6 marks)*

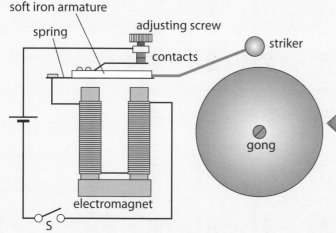

When the switch is closed current flows through the coils of the electromagnet. A magnetic field is produced. The soft iron armature is attracted to the electromagnet. This makes the striker move and hit the gong. When the armature moves, the contacts move apart, breaking the circuit so the current stops flowing. The armature is not attracted to the electromagnet and the spring pulls it back to its starting position. This closes the contacts. If the switch is still closed, the whole sequence repeats over and over again.

Suggest

The command question here is SUGGEST. That means that you may not have studied this exact example but you should be able to apply your knowledge of electromagnets to the question.

Make sure that you use the artwork you are given to help you.

This is a good answer because it describes in detail what might happen at every stage of the process.

Now try this

A transformer used for a model railway has a sticker on it.

Explain the structure and function of this transformer and the effect it has on the current in the model train.   *(6 marks)*

You should include some information on why the transformer is necessary for the model train

Warning: mains electricity
Input: 230 V a.c.
Output: 11.5 V
Maximum current: 3 A

Periodic Table

PERIODIC TABLE

Key:

| relative atomic mass |
| **atomic symbol** |
| name |
| atomic (proton) number |

| 1 | | hydrogen | 1 |

| | | | | | | | |

Group 1

| 7
Li
lithium
3 |
| 23
Na
sodium
11 |
| 39
K
potassium
19 |
| 85
Rb
rubidium
37 |
| 133
Cs
caesium
55 |
| 223
Fr
francium
87 |

Group 2

| 9
Be
beryllium
4 |
| 24
Mg
magnesium
12 |
| 40
Ca
calcium
20 |
| 88
Sr
strontium
38 |
| 137
Ba
barium
56 |
| 226
Ra
radium
88 |

Transition elements

45 Sc scandium 21	48 Ti titanium 22	51 V vanadium 23	52 Cr chromium 24	55 Mn manganese 25	56 Fe iron 26	59 Co cobalt 27	59 Ni nickel 28	63.5 Cu copper 29	65 Zn zinc 30
89 Y yttrium 39	91 Zr zirconium 40	93 Nb niobium 41	96 Mo molybdenum 42	99 Tc technetium 43	101 Ru ruthenium 44	103 Rh rhodium 45	106 Pd palladium 46	108 Ag silver 47	112 Cd cadmium 48
139 La lanthanum 57	178 Hf hafnium 72	181 Ta tantalum 73	184 W tungsten 74	186 Re rhenium 75	190 Os osmium 76	192 Ir iridium 77	195 Pt platinum 78	197 Au gold 79	201 Hg mercury 80
227 Ac actinium 89	261 Rf rutherfordium 104	262 Db dubnium 105	266 Sg seaborgium 106	264 Bh bohrium 107	277 Hs hassium 108	268 Mt meitnerium 109	271 Ds darmstadtium 110	272 Rg roentgenium 111	

Groups 3, 4, 5, 6, 7, 0

3	4	5	6	7	0
					4 **He** helium 2
11 **B** boron 5	12 **C** carbon 6	14 **N** nitrogen 7	16 **O** oxygen 8	19 **F** fluorine 9	20 **Ne** neon 10
27 **Al** aluminium 13	28 **Si** silicon 14	31 **P** phosphorus 15	32 **S** sulfur 16	35.5 **Cl** chlorine 17	40 **Ar** argon 18
70 **Ga** gallium 31	73 **Ge** germanium 32	75 **As** arsenic 33	79 **Se** selenium 34	80 **Br** bromine 35	84 **Kr** krypton 36
115 **In** indium 49	119 **Sn** tin 50	122 **Sb** antimony 51	128 **Te** tellurium 52	127 **I** iodine 53	131 **Xe** xenon 54
204 **Tl** thallium 81	207 **Pb** lead 82	209 **Bi** bismuth 83	210 **Po** polonium 84	211 **At** astatine 85	222 **Rn** radon 86

The lanthanides (atomic numbers 58–71) and the actinides (atomic numbers 90–103) have been omitted.

Elements with atomic numbers 112–118 have been reported but not fully authenticated.

Cu and Cl have not been rounded to the nearest number.

Chemistry Data Sheet

Reactivity series of metals

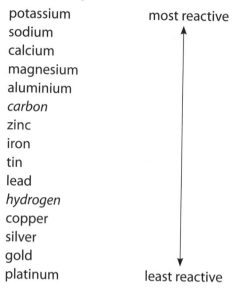

potassium   most reactive
sodium
calcium
magnesium
aluminium
carbon
zinc
iron
tin
lead
hydrogen
copper
silver
gold
platinum   least reactive

Elements in italics, though non-metals, have been included for comparison.

Formulae of some common ions

Positive ions

Name	Formula
hydrogen	H^+
sodium	Na^+
silver	Ag^+
potassium	K^+
lithium	Li^+
ammonium	NH_4^+
barium	Ba^{2+}
calcium	Ca^{2+}
copper(II)	Cu^{2+}
magnesium	Mg^{2+}
zinc	Zn^{2+}
lead	Pb^{2+}
iron(II)	Fe^{2+}
iron(III)	Fe^{3+}
aluminium	Al^{3+}

Negative ions

Name	Formula
chloride	Cl^-
bromide	Br^-
fluoride	F^-
iodide	I^-
hydroxide	OH^-
nitrate	NO_3^-
oxide	O^{2-}
sulfide	S^{2-}
sulfate	SO_4^{2-}
carbonate	CO_3^{2-}

Physics Equations Sheet

$s = v \times t$	s distance v speed t time
refractive index $= \dfrac{\sin i}{\sin r}$	i angle of incidence r angle of refraction
magnification $= \dfrac{\text{image height}}{\text{object height}}$	
$P = \dfrac{1}{f}$	P power f focal length
refractive index $= \dfrac{1}{\sin c}$	c critical angle
$T = \dfrac{1}{f}$	T periodic time f frequency
$M = F \times d$	M moment of the force F force d perpendicular distance from the line of action of the force to the pivot
$P = \dfrac{F}{A}$	P pressure F force A cross-sectional area
$\dfrac{V_p}{V_s} = \dfrac{n_p}{n_s}$	V_p potential difference across the primary coil V_s potential difference across the secondary coil n_p number of turns on the primary coil n_s number of turns on the secondary coil
$V_p \times I_p = V_s \times I_s$	V_p potential difference across the primary coil I_p current in the primary coil V_s potential difference across the secondary coil I_s current in the secondary coil

AQA specification skills

In your AQA exam there are certain **skills** that you sometimes need to **apply** when answering a question. Questions often contain a particular **command word** that lets you know this. On this page we explain how to spot a command word and how to apply the required skill.

> Note: Watch out for our Skills sticker – this points out the Worked Examples that are particularly focused on applying skills.

Command word	Skill you are being asked to apply
Compare Page 79	**Compare** how two things are similar or different. Make sure you include both of the things you are being asked to compare. For example: 'A is bigger than B, but B is lighter than A.'
Consider Page 79	You will be given some information and you will be asked to **consider** all the factors that might influence a decision. For example: 'When buying a new fridge the family would need to consider the following things …'
Describe Page 79	**Describe** a process or why something happens in an accurate way. For example: 'When coal is burned the heat energy is used to turn water into steam. The steam is then used to turn a turbine …'
Discuss Page 79	In some questions you might be asked to make an informed judgement about a topic. This might be something like stem cell research. You should **discuss** the topic and give your **opinion** but make sure that you back it up with information from the question or your scientific knowledge.
Draw Page 79	Some questions ask you to **draw** or sketch something. It might be the electrons in an atom, a graph or a ray diagram. Make sure you take a pencil, rubber and ruler into your exam.
Evaluate Page 79	This is the most important one! Most of the skill statements start with **evaluate**. You will be given information and will be expected to use that information plus anything you know from studying the material in the specification to look at evidence and come to a **conclusion**. For example, if you were asked to evaluate which of two slimming programmes was better, then you might comment like this: 'In programme A people lost weight quickly to start with but then put the weight back on by the end of the sixth month. In programme B they did not lose weight so quickly to start with, but the weight loss was slow and steady and no weight was gained back by the end of the year. I therefore think that programme B is most effective.'
Explain Page 79	State what is happening and **explain** why it is happening. If a question asks you to explain then it is a good idea to try to use the word 'because' in your answer. For example: 'pH 2 is the optimal pH for enzymes in the stomach because the stomach is very acidic.'
Interpret Page 79	**Interpret** the data given to you on graphs, diagrams or in tables to help answer the question. For example: 'Use the data to show what happens when …'
Suggest Page 79	You will be given some information about an unfamiliar situation and asked to **suggest** an answer to a question. You will not have learned the answer – you need to **apply** your knowledge to that new situation. For example: 'I think that blue is better than green because …' or 'It might be because …'

ISA support

25% of each Science GCSE comes from controlled assessments called ISAs (Investigative Skills Assignments). These have four stages:

1 Carrying out some research to plan an experiment to investigate a hypothesis.
2 ISA Paper 1: a 45-minute question paper about your experiment plan done under exam conditions.
3 Carrying out your experiment (or a similar one with instructions provided by your teacher).
4 ISA Paper 2: a 50-minute question paper about your results and some similar case studies done under exam conditions.

All the examples on this page and the next are from the same investigation. The investigation looked at the temperature rise measured when different masses of zinc powder are added to copper sulfate solution. Here is the equation for the reaction that takes place:

$$CuSO_4(aq) + Zn(s) \rightarrow Cu(s) + ZnSO_4(aq)$$

Worked example

Hypothesis (Paper 1)

Write a hypothesis about how the mass of zinc used affects the temperature rise in the reaction between copper sulfate solution and zinc powder. Explain why you made this hypothesis.

Hypothesis – I predict that the temperature rise will increase as more zinc is reacted with copper sulfate solution.

Explanation – The reaction is exothermic, which means that it releases heat energy, which increases the temperature. The more zinc that is used, the more bonds that are made as the zinc reacts with the copper sulfate, and so the bigger the temperature rise.

> The hypothesis is very sensible and is justified with some good science. The student has clearly explained why <u>increasing</u> the mass of zinc will <u>increase</u> the temperature rise, not just why changing the mass of zinc will change the temperature rise. The student explains clearly why the temperature rises.

> Higher tier pupils could also discuss bond energies. For example, that more energy is released when new bonds are formed than is used when old bonds are broken.

Research sources (Paper 1)

Name two sources you used for your research. Which of the two sources did you find most useful and why?

My sources were:

1 AQA GCSE Chemistry (Longman, editor Nigel English)

2 http://www.nuffieldfoundation.org/practical-chemistry

Source 1 was useful because it explained the scientific theory of what happens in exothermic reactions. Source 2 was even more useful because it showed me a simple method I could use to measure the temperature rise in a reaction.

> This is a very good answer as it gives the title, author and publisher of the book and full URL of the website. It also gives reasons for why each source was useful.

Results table (Paper 1)

Draw a table to record the results of your investigation.

Mass of zinc in g	Start temperature in °C	Final temperature in °C	Temperature rise in °C
1			
2			
3			
4			
5			

> This table is very good as it contains columns for all the measurements that need to be taken. This includes the start and end temperatures rather than just the temperature rise. It also has clear column headings and includes units.

ISA support

Worked example

Plan (Paper 1)

Describe how you plan to do your investigation to test the hypothesis given. You should include:

- the equipment you plan to use
- how you will use the equipment
- the measurements you are going to make
- how you will make the investigation a fair test
- a risk assessment.

Variables

independent variable = mass of zinc

dependent variable = temperature rise

control variables = volume of copper sulfate solution, concentration of copper sulfate solution, surface area of zinc

> It is a good idea to state clearly what the independent, dependent and control variables are at the start. This will help you to be clear what you are trying to do.

Method

1 I will place 25 cm³ of copper sulfate solution into a polystyrene cup. I will measure the volume with a measuring cylinder.

2 I will place a lid on the cup and place it in a beaker to stop the cup falling over. I will also place the beaker and cup under a tripod to support the thermometer.

3 I will measure the starting temperature of the copper sulfate solution.

4 I will weigh out 1 g of zinc powder using a balance.

5 I will add the zinc to the copper sulfate solution and stir well with the thermometer. I will record the highest temperature reached.

6 I will repeat the experiment with 2, 3, 4 and 5 g of powdered zinc.

thermometer

tripod to hold thermometer

beaker to support cup

cup with lid

reaction mixture

> Make sure that you label your diagram.

> Your instructions need to be clear. A good guide is to write the instructions so that someone else in your class could follow them to do your experiment. They need to be written in good English, using full sentences, but they can be numbered. A diagram can be helpful. Use scientific language where it is appropriate.

Apparatus

Polystyrene cup, lid, beaker, thermometer, tripod, balance, zinc powder, copper sulfate solution

> It helps to list all the apparatus that you use.

Fair testing

In order to make this a fair test I must use the same volume and concentration of copper sulfate solution, as well as using zinc with the same surface area each time. I will use zinc powder from the same supplier to ensure this is kept the same.

> You need to state clearly how you will make the experiment a fair test. You must state which variables you will keep the same.

Risk assessment:

This is a medium-risk experiment. The main dangers in this experiment are that copper sulfate is harmful and zinc powder is flammable. I will wear eye protection throughout, ensure there are no naked flames in the lab and wash my hands after the experiment.

> You need to identify real risks and suggest ways to prevent these problems happening.

ISA support

Paper 2 of the ISA looks at your own results and those from other similar case studies. In many of the questions, it is important to remember to use the data to justify your answers.

The next few examples relate to an investigation into pendulums. A company makes 'grandfather clocks' that use pendulums. They wish to investigate the factors that affect the time period of a pendulum.

The examples on this page are from an experiment where a student investigated how changing the length of a pendulum affected the time it took to complete one swing. The student's hypothesis was: The longer the pendulum, the longer it takes to complete one swing.

Here are the results recorded by the student.

This is the equipment the students used:

Length of pendulum in cm	Time for 10 swings in seconds				Average time for 1 swing in seconds
	1	2	3	Ave	
10	7	6	6	6	0.6
20	9	8	9	9	0.9
30	11	10	12	11	1.1
40	12	13	13	13	1.3
50	14	14	15	14	1.4

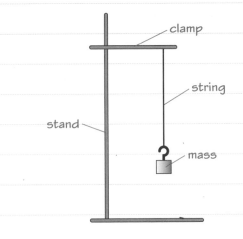

Worked example

Conclusions

Do your results support your hypothesis? Give some evidence to support your conclusion.

Yes the results support my hypothesis. The longer the pendulum, the longer the time taken for one swing. For example, when the pendulum is 10 cm long each swing takes 0.6 seconds, but when it is 50 cm long it takes 1.4 seconds.

This is a very good answer. First, it clearly states whether the results support the hypothesis. Second, it states what the hypothesis is. Finally, it quotes some data from the results to justify this conclusion.

ISA support

Worked example

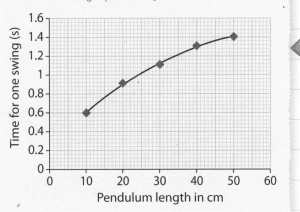

Here is the graph of my results.

You will be awarded up to 4 marks for your chart or graph.

You will always have to plot a bar chart or line graph. Plot a bar chart for categoric variables (those that are described in words, e.g. colours). Plot a line graph for continuous variables (those that are described by numbers, e.g. length). The independent variable should be on the horizontal axis with the dependent variable on the vertical axis.

This is a good graph. The axes have suitable scales. Both axes are labeled and have units. A good best fit line has been drawn that is a smooth and has a similar number of points above and below the line. Best fit lines can be straight or curved.

Here is the data from four Case Studies that were carried out. In the exam, these will be given to you on a Secondary Data Sheet.

Worked example

Case study 1

A student timed how long it took for a pendulum of different lengths to complete 50 swings.

Length of pendulum in cm	Time for 50 swings in s
20	44
40	63
60	77
80	90
100	100

Case study 2

A student timed how long it took a pendulum with different masses on the end to complete 10 swings.

Mass on pendulum in g	Time for 10 swings in s
100	12
200	13
300	12
400	12

Case study 3

A student timed how long it took for another pendulum of different lengths to complete 1 swing.

Length of pendulum in cm	Time for 1 swing in s		
	Test 1	Test 2	Mean
10	3	4	3.5
20	4	4	4.0
30	5	6	5.5
40	6	7	6.5
50	7	7	7.0

Case study 4

A company making swings for playgrounds tested swings of different lengths to see what effect it had on the time for each swing for two children of different masses on the swing.

Length of swing in m	Mass of child in kg	Time for one swing in s
1.5	30	2.4
1.5	50	2.5
2.0	30	2.8
2.0	50	2.8
2.5	30	3.2
2.5	50	3.1

ISA support

Worked example

Using case studies (Paper 2)

Explain whether the results from Case Studies 1, 2 and 3 on the Secondary Data Sheet support your hypothesis. To gain full marks, your explanation should include appropriate examples from the results.

Case study 1 supports my hypothesis. As the pendulum increases in length from 20 to 100 cm the time it takes for 50 swings increases from 44 to 100 seconds. Case study 2 is irrelevant as the mass on the end of the pendulum was changed rather than the length. Case study 3 supports my hypothesis. As the pendulum increases in length from 10 to 50 cm the time it takes for 1 swing increases from 3.5 to 7.0 seconds.

This answer is good because it clearly states whether all three case studies support the hypothesis and quotes data (including units) to justify the answer. For Case Study 2 it clearly explains why it is irrelevant, as a different variable has been changed.

Using case studies (Paper 2)

Explain whether the results from Case Study 4 support your hypothesis. To gain full marks your explanation should include appropriate examples from the results.

Case study 4 does support my hypothesis. If the results for the 30 kg child are used, then time for one swing increases from 2.4 to 3.2 seconds as the swing increases in length from 1.5 to 2.5 m. Similarly, if the results for the 50 kg child are used, then time for one swing increases from 2.5 to 3.1 seconds as the swing increases in length from 1.5 to 2.5 m.

This answer is good because it picks out the relevant data to show that it does support the hypothesis. It actually picks out two sets of data to support it. The answer also contains the units.

Relating to context (Paper 2)

Explain how the results of your experiment could be useful in the context you have been given.

The results show how the length of the pendulum affects the time for a complete swing of the pendulum. In a grandfather clock the length of the pendulum controls the time, with the time period for one complete swing often being 2 seconds. If the clock is going too fast, then the pendulum must be swinging too fast. These results show that the weight should be moved further down to make the pendulum longer and increase the time period. If the clock is going too slow, the weight should be moved up to decrease the length and the time for one swing.

This answer is very good for several reasons. The student has researched the context they were given to see how the pendulum controls the time. The student then clearly uses their results to show how the weight on the pendulum can be moved to change the length of the pendulum to change the time period.

Answers

Biology answers

1. Into and out of cells

1. Osmosis is the net movement of water molecules from a region of high concentration to a region of lower concentration (of water) **(1)** across a partially permeable membrane **(1)**.

2. **(a)** Neither diffusion nor osmosis need energy **(1)**.

   (b) Diffusion does not need energy but active transport does need energy/diffusion takes place down a concentration gradient, active transport takes place against a concentration gradient **(1)**.

3. The cells will still be able to absorb water **(1)** by osmosis because that is a passive process **(1)**. The cells will not be able to absorb mineral ions **(1)**; mineral ions are absorbed by active transport and that needs energy from respiration **(1)**.

2. Sports drinks

1. in sweat **(1)**

2. The drink replaces sugars that have been used in respiration **(1)** so that the body can continue to release energy at a higher rate for longer **(1)**.

3. If the runner drinks the hypotonic drink, it will replace water lost in sweat, sugar used in respiration and a few of the ions lost in sweat **(1)**. But the concentration of ions will decrease in the body, which may reduce how well the cells work **(1)**. An isotonic drink replaces water *and* ions in the right balance so there is no change in water/ion balance in the body during the race **(1)**.

3. Exchanging materials

1. Any one from: lungs, small intestine **(1)**. *There are other possible answers, e.g. kidney, but the lungs and small intestine are the ones that you will be expected to know.*

2. Any one answer from: they increase the surface area of the lungs; they are thin-walled and so provide a short distance for diffusion **(1)**.

3. The capillaries deliver blood that has high carbon dioxide levels and low oxygen levels and takes away blood that has high oxygen levels and low carbon dioxide levels **(1)**. This means that there is always a concentration gradient between the blood and the air in the lungs **(1)**. Ventilation of the lungs replaces air with higher concentration of carbon dioxide and lower concentration of oxygen with air that has a higher concentration of oxygen and lower concentration of carbon dioxide **(1)**. These factors maintain a high concentration gradient between the body and air for the two gases **(1)**.

4. Ventilation

1. They move out and up **(1)**.

2. During breathing in **(1)**, because the muscles between the ribs contract to move the ribs up and out **(1)** and the muscle surrounding the diaphragm contracts to flatten the diaphragm **(1)**.

3. Increased pressure on the lungs would decrease their volume **(1)**. The pressure inside the lungs would be greater than the outside air pressure, so air would be forced out of the lungs through the mouth and/or nose **(1)**. Decreasing the pressure inside the case would decrease the pressure on the lungs and so increase their volume **(1)**. This would create a lower pressure inside the lungs than air pressure, so air would be drawn into the lungs through the mouth and/or nose **(1)**.

5. Exchange in plants

1. They have internal air spaces **(1)** that increase the surface area for diffusion of gases **(1)**. They have stomata in the leaf surface **(1)** so that gases can quickly enter and leave the leaf **(1)**.

2. **(a)** Water evaporates from leaves more quickly when the plant is in warm conditions **(1)**. The plant near the radiator was probably not able to absorb water quickly enough through its roots to replace the water that evaporated, so it wilted **(1)**.

   (b) The stomata would be closed **(1)** to prevent further water loss **(1)**.

   (c) Guard cells either side of the stoma change shape, either creating or closing the gap between them **(1)**.

6. The circulatory system

1. left atrium **(1)**

2. They prevent the blood flowing the wrong way in the heart **(1)**.

3. The right ventricle pumps blood to the lungs, which are not far away **(1)**. The left ventricle pumps blood out around the rest of the body, including as far as the toes and fingers **(1)**. So the left ventricle needs to pump with greater force than the right ventricle **(1)**.

7. Blood vessels

1. **(a)** capillary   **(b)** artery

2. The stent increases the width of the artery **(1)** allowing more blood to flow through to the tissue beyond the narrowed part **(1)**.

3. Every body cell needs oxygen and glucose for respiration, and to get rid of waste products such as carbon dioxide **(1)**. If this doesn't happen quickly enough then the cell may be damaged **(1)**. Capillaries are the blood vessels that exchange substances with the body cells **(1)**. A short distance for diffusion between a body cell and a capillary helps to make diffusion rapid **(1)**.

8. Blood

1. To help protect the body against infection by microorganisms **(1)**.

2. The chemicals in artificial blood are similar to haemoglobin in red blood cells **(1)**, which combines with oxygen in the lungs **(1)** and releases the oxygen when in capillaries that run through organs **(1)**.

3. Any three suitable statements, such as: during combat, many soldiers may receive large wounds that bleed a lot and could result in death, so more blood might be needed than is available **(1)**; soldiers may be hurt in areas where normal blood cannot be stored properly **(1)**; artificial blood can be given to any soldier, but human blood can only be given to soldiers who have the same type of blood **(1)**.

9. Transport in plants

1. xylem **(1)**, phloem **(1)**

2. From the phloem **(1)**, because dissolved sugars are transported around the plant in phloem **(1)**.

3. Sugars are made in the leaves after photosynthesis **(1)**. In the spring and summer most of the sugars will be transported in phloem to the growing parts of the plant, such as the tips of shoots and roots **(1)**. In the autumn, less sugar will transported to these growing parts and more will be transported to the potatoes for storage **(1)**.

10. Biology six mark question 1

Answers can be found on page 91.

11. Removing waste products

1. **(a)** liver **(1)**   **(b)** kidney **(1)**

2. Filtering **(1)**; removes some substances from the blood **(1)**. Reabsorption **(1)** of sugars, water and dissolved ions needed by the body **(1)**.

3. Active transport moves substances against their concentration gradient **(1)**. The concentration of sugar and ions is greater in the blood than inside the kidney, so they can only be reabsorbed by active transport **(1)**. Osmosis is a passive process, in which water moves down its concentration gradient across a partially permeable membrane **(1)**. The concentration of water in the blood is less than inside the kidney, so water can be reabsorbed by osmosis **(1)**.

12. Kidney treatments

1. dialysis **(1)**, kidney transplant **(1)**

2. The blood flows in a tube with a partially permeable membrane **(1)** through dialysis fluid **(1)**. Dissolved substances in the fluid and the blood are exchanged **(1)**. The substances in the dialysis fluid are at the right concentrations to restore their normal levels in the blood **(1)**.

3. All cells have surface proteins called antigens **(1)**. A transplanted kidney comes from a different person so will have different antigens **(1)**. The antibodies of the immune system of the person given the kidney will attack the transplanted kidney and cause the body to reject it **(1)**.

13. Body temperature

1. The thermoregulatory centre **(1)**. *An answer of 'the brain' would not be enough.*

2. When it is hot we sweat more **(1)**. We need to drink more to replace the water lost in sweat **(1)**.

3. Reduced blood flow near the skin surface **(1)** means that less heat is transferred to the environment through the skin **(1)**. Shivering is contraction of the muscles, **(1)** which uses energy from respiration and releases some energy as heat, which warms the body **(1)**.

14. Blood glucose control

1. Insulin decreases the concentration of glucose in the blood **(1)** and glucagon increases the concentration of glucose in the blood **(1)**.

2. Different amounts of glucose will be absorbed from different kinds of food **(1)**. The amount of insulin injected must match the amount of glucose that is absorbed from a meal, **(1)** to make sure that blood glucose concentration doesn't get too high or fall too low **(1)**.

15. Biology six mark question 2

Answers can be found on page 91.

16. Pollution

1. Any one from: sewage/human waste, fertiliser, toxic chemicals/pesticides and herbicides washed in from the land **(1)**.

2. Any two effects with suitable explanations, such as: destruction of habitats for the plants and animals that live there now **(1)** as the land is dug up to get the building materials **(1)**; more air pollution in the area **(1)** caused by all the machinery and vehicles needed to dig up, process and remove the building materials **(1)**.

3. More people need more food so there is more farming **(1)**. More farming means more use of chemicals such as pesticides and herbicides, which means more pollution **(1)**.

17. Deforestation

1. Any one from: for timber, to clear land for growing crops, to clear land for cattle to graze **(1)**.

2. This will reduce the destruction of peat bogs **(1)** and reduce carbon dioxide released into the air from their destruction **(1)**.

3. Clearing land by burning trees releases carbon dioxide into the air **(1)**; increased activity of microorganisms that decay dead wood releases more carbon dioxide from respiration **(1)**; carbon dioxide may be removed from the atmosphere more slowly or it may not be 'locked up' in wood for as long. **(1)**

18. Global warming

1. Any one from: cause more flooding, more drought, make it hotter, make it colder in some places **(1)**.

2. The organisms that cause tropical diseases, or the organisms that transmit the diseases to people, can only live where it is warm **(1)**. If the climate gets warmer in other places, then these organisms may be able to survive there and so people in those places may catch the diseases **(1)**.

3. Water in oceans absorbs a lot of the carbon dioxide from the air **(1)**. In the oceans the carbon dioxide is sequestered and removed from the carbon cycle **(1)**. If less carbon dioxide is absorbed and sequestered, there will be more carbon dioxide in the air **(1)**. This could cause even more global warming **(1)**.

19. Biofuels

1. anaerobic **(1)**; methane **(1)**

2. Any two from: wide range of waste materials from agriculture for commercial biogas production, animal and human waste used in domestic biogas production **(1)**; leftovers sold as fertiliser in commercial biogas, used on garden crops in domestic biogas **(1)**; biogas generates electricity in commercial biogas, biogas burnt for heating and lighting in domestic biogas **(1)**.

3. Burning fossil fuels releases carbon that was locked away in the fuels for millions of years **(1)** and so rapidly adds much more carbon dioxide to the air **(1)**. Burning biogas releases carbon that was recently taken from the air by plants and animals **(1)**. So the change in carbon dioxide over a short time averages out as zero **(1)**.

20. Food production

1. Restricting movement reduces the amount of energy the animal uses for muscle contraction **(1)**. Increasing the air temperature reduces the amount of heat energy transferred from the animal to the environment **(1)**. This means that more energy from food goes to making new animal tissue **(1)**.

2. Usually the lowest level of a food chain is the most efficient for gathering food for humans **(1)** because energy is lost to the environment at each level of the food chain **(1)**. In this food chain it might be difficult to collect large enough amounts of plankton from the sea to make into food **(1)**. So catching and eating sardines will be more efficient in terms of food production than eating tuna, but might also be more efficient than eating plankton **(1)**.

21. Fishing

1. Overfishing may have taken all the fish that we eat **(1)**.

2. Fishing quotas can stop fishing taking too many fish from an area **(1)**. Net sizes can make sure some breeding fish are always left in the area **(1)**.

3. Any suitable answer that shows a conflict between the way scientists base their advice on measurements and the way politicians consider how this will affect people; for example: Scientists will base their advice on measurements and predictions for the future **(1)**, but the people who set the targets try to make sure they don't hurt fishing communities too much/ that there are plenty of fish in the shops for people to buy **(1)**.

22. Sustainable food

1. The distance food is transported from where it is grown to where it is sold **(1)**.

2. Vegetarians don't eat meat **(1)**. Mycoprotein is made from a fungus and not from animals **(1)**.

3. Any four suitable conditions, such as: The fungus is aerobic, so needs air if it is to grow well **(1)**; glucose syrup is added to provide energy for respiration **(1)**; temperature should be controlled to keep the fungus warm but not too hot **(1)**; other nutrients are needed to make new cells **(1)**.

23. Biology six mark question 3

Answers can be found on page 91–2.

Chemistry answers

24. The early periodic table

1. (a) Group 0/the noble gases **(1)**

    (b) Similarities: both groups contain lithium, sodium, potassium and rubidium **(1)**, these metals are given in the same order **(1)**. Difference: the modern Group 1 does not contain hydrogen/copper/silver **(1)**.

2. Any two from the following: there were gaps **(1)**; some groups contained metals and non-metals **(1)**; Mendeleev swapped iodine and tellurium to suit their properties rather than their atomic weight **(1)**; the boxes in lower rows contained two elements rather than one **(1)**.

25. The modern table

1. Any two from the following: atoms in the same group have the same number of electrons in their outer shells **(1)**; apart from Group 0 the number of electrons in the outer shell is the same as the group number **(1)**; the number of occupied shells is the same as the period number **(1)**; a shell is being filled as you go across a period **(1)**.

2. Mendeleev put tellurium first so that iodine would be in the same group as fluorine/chlorine/bromine **(1)**. The modern table is based on atomic number/electronic structure **(1)** so tellurium should come first **(1)**.

26. Group 1

1. Any three from the following: they are all metals **(1)**; have a low density **(1)**; form ions with a charge of +1 **(1)**; react with non-metals to form ionic compounds **(1)**; react with water to form hydrogen/alkaline hydroxides **(1)**.

2. Any stated temperature above 20 °C but below 98 °C **(1)**; because the melting point decreases down the group and potassium is below sodium **(1)**, and potassium is a solid at room temperature so its melting point must be above 20 °C **(1)**.

3. (a) lithium + water → lithium hydroxide + hydrogen **(1)**

    (b) LiOH **(1)**

4. (a) Group 1 elements are metals that react with water to form alkaline solutions **(1)**.

    (b) $2Rb(s) + 2H_2O(l) \rightarrow 2RbOH(aq) + H_2(g)$ (**1 mark** for balancing, **1 mark** for state symbols)

27. Transition metals

1. Three from the following: form coloured compounds **(1)**; useful as catalysts **(1)**; form more than one type of ion **(1)**; (compared to Group 1) denser/harder/stronger/less reactive/ higher melting points/higher boiling points **(1)**.

2. Sodium is in Group 1 and forms colourless/white compounds **(1)**; copper is a transition metal and forms coloured compounds **(1)**. So the white powder is sodium oxide and the red powder is copper(I) oxide **(1)**.

3. Any three from the following: copper is stronger **(1)**; harder; less reactive **(1)**; has a higher melting point **(1)**.

4. Iron can act as catalyst **(1)** because it is a transition metal **(1)**.

28. Group 7

1. Any two from the following: non-metals **(1)**; form ions with a charge of–1 **(1)**; react with metals to form ionic compounds **(1)**.

2. Astatine should be a solid **(1)** because the melting points and boiling points increase down the group **(1)** (and astatine is below iodine, which is a solid).

3. (a) The mixture turns brown **(1)**.

    (b) This is because bromine is more reactive than iodine **(1)**, so it displaces iodine from potassium iodide solution **(1)**. bromine + potassium iodide → potassium bromide + iodine **(1)**.

4. Chlorine atoms are larger than fluorine atoms **(1)**; the outer shell is further from the nucleus **(1)**, so chlorine gains electrons less easily **(1)**.

29. Hard and soft water

1. Any two from the following: same volume of water **(1)**; same volume of soap solution **(1)**; same concentration of soap solution **(1)**.

2. Temporary hard water contains hydrogencarbonate ions **(1)**, which decompose when the water is heated **(1)** to form carbonate ions **(1)** that form a precipitate with magnesium ions **(1)**. $Mg^{2+}(aq) + CO_3^{2-}(aq) \rightarrow MgCO_3(s)$ **(1)**.

30. Softening hard water

1. Benefit – one from: good for the teeth and bones; helps to reduce heart disease **(1)**. Problem – one from: more soap is needed; efficiency of heating systems and kettles is reduced **(1)**.

2. Sodium carbonate removes calcium ions (and/or magnesium ions) **(1)** by forming a precipitate of calcium carbonate (and/or magnesium carbonate) **(1)**.

3. Ion exchange resin swaps calcium and/or magnesium ions for sodium ions **(1)** so sodium ions are used up over time **(1)**; sodium chloride is added to replace the missing sodium ions **(1)**.

31. Purifying water

1. filter beds to remove solids/leaves/sticks **(1)**; chlorine to sterilise the water/reduce microbes **(1)**

2. Silver is too expensive to use in large amounts **(1)**; chlorine is dangerous so cannot be used at home **(1)**.

3. Reason in favour of using chlorine, for example, chlorine kills harmful microbes **(1)**; reason against using it, for example, it is toxic or spoils the taste or smell of the water **(1)**; conclusion, for example, without chlorine there would be more disease so it is worth adding **(1)**.

32. Chemistry six mark question 1

Answers can be found on page 92.

33. Calorimetry

1. $Q = 50 \times 4.2 \times 8$ **(1)** $= 1680$ J **(1)** *The total mass of solution is 25 + 25 g.*

2. 240×4.2 **(1)** $= 1008$ J/g **(1)**

ANSWERS

34. Energy level diagrams

1. The energy level of the reactants is more than the energy level of the products **(1)** and overall energy is given out **(1)**.

2. (a) hydrogen + oxygen → water **(1)**

   (b) The flame overcomes the activation energy/provides the energy needed to start the reaction **(1)**.

3. The line goes up when energy is supplied to break bonds in the reactants **(1)**; this is the activation energy **(1)**; it goes down again when energy is released when bonds form in the reactants **(1)**; the level of the products is less than the level of the reactants because it is an exothermic reaction/energy is released to the surroundings **(1)**.

35. Bond energies

1. (a) Energy in to break bonds: $4 \times (O-H) = 4 \times 464 = 1856$ kJ **(1)**. Energy out when new bonds form: $2 \times (H-H) = 2 \times 436 = 872$ kJ, $1 \times (O=O) = 1 \times 498 = 498$ kJ **(1)**; total $= 872 + 498 = 1370$ kJ **(1)**. Overall energy change $= 1856 - 1370 = 486$ kJ **(1)**

   (b) The process is endothermic **(1)**, because the energy needed to break existing bonds is greater than the energy released from forming new bonds **(1)**.

2. Energy in to break bonds: $4 \times (C-H) = 4 \times 413 = 1652$ kJ and $2 \times (O=O) = 2 \times 498 = 996$ kJ **(1)**; total $= 1652 + 996 = 2648$ kJ **(1)**. Energy out when new bonds form: $4 \times (O-H) = 4 \times 464 = 1856$ kJ and $2 \times (C=O) = 2 \times 805 = 1610$ kJ **(1)**; total $= 1856 + 1610 = 3466$ kJ **(1)**. Overall energy change $= 2648 - 3466 = -818$ kJ **(1)**

36. Hydrogen as a fuel

1. hydrogen + oxygen → water **(1)**

2. (a) One advantage, e.g. they will make the environment more peaceful/busy roads will not be so noisy **(1)**. One disadvantage, e.g. it will be difficult for people to hear them coming/there may be more pedestrian accidents **(1)**.

   (b) One sensible suggestion, e.g. they use platinum/platinum is expensive/they are still under development **(1)**.

3. Any two advantages of hydrogen as a fuel from, for example: hydrogen does not produce carbon dioxide when it is burned (but hydrocarbons do) **(1)**; hydrogen does not produce carbon monoxide (but hydrocarbons do) **(1)**; hydrogen produces fewer greenhouse gases than hydrocarbons **(1)**; hydrogen can be made from renewable resources but hydrocarbons are fossil fuels **(1)**. Any two disadvantages of hydrogen as a fuel, for example: hydrogen is difficult to store (but hydrocarbons are easy to store and transport) **(1)**; there are only a few filling stations that sell hydrogen (hydrocarbons are widely available **(1)**. *Note: Neutral statements are valid, for example: both fuels produce NO_x, and the engines are as noisy as each other.*

37. Tests for metal ions

1. Add sodium hydroxide solution **(1)**; iron(II) chloride forms a green precipitate **(1)** and iron(III) chloride forms a brown precipitate **(1)**.

2. (a) Red flame **(1)** in a flame test **(1)**; white precipitate **(1)** with sodium hydroxide solution **(1)**.

   (b) Magnesium and calcium both produce white precipitates with sodium hydroxide solution **(1)** and neither dissolves in excess sodium hydroxide solution **(1)**.

3. The different colours in the flame test will mix or interfere with each other **(1)**.

38. More tests for ions

1. (a) bubbles/fizzing/effervescence **(1)**

   (b) Pass the gas through limewater **(1)**, which will turn cloudy **(1)**.

   (c) The yellow flame test shows that sodium ions are present **(1)**, so the washing soda must be sodium carbonate **(1)**.

2. Seawater and the damp both contained chloride ions **(1)**, but the rainwater did not **(1)**, so the damp probably came from seawater **(1)**.

3. Flame test: potassium iodide and potassium bromide both give a lilac flame **(1)**. Add dilute nitric acid then silver nitrate solution to these two **(1)**. Potassium iodide gives a yellow precipitate **(1)** and potassium bromide gives a cream precipitate **(1)**. Add dilute sodium hydroxide to the other two – both give a white precipitate **(1)**, but only aluminium sulfate will give a precipitate that dissolves in excess sodium hydroxide **(1)**.

39. Titration

1. volume of potassium hydroxide $= 25.0 \div 1000 = 0.025$ dm³; moles of potassium hydroxide $= 0.20 \times 0.025$ **(1)** $= 0.005$ mol **(1)**; volume of hydrochloric acid $= 20.0 \div 1000 = 0.02$ dm³; concentration of hydrochloric acid $= 0.005 \div 0.02$ **(1)** $= 0.25$ mol/dm³ **(1)**

2. (a) volume of hydrochloric acid $= 24.0 \div 1000 = 0.024$ dm³; moles of hydrochloric acid $= 0.50 \times 0.024$ **(1)** $= 0.012$ mol **(1)**; volume of sodium hydroxide $= 25.0 \div 1000 = 0.025$ dm³; concentration of sodium hydroxide $= 0.012 \div 0.025$ **(1)** $= 0.48$ mol/dm³ **(1)**

   (b) concentration in g/dm³ $= M_r \times$ concentration in mol/dm³ $= 40 \times 0.48$ **(1)** $= 19.2$ g/dm³ **(1)**

40. Chemistry six mark question 2

Answers can be found on page 92.

41. The Haber process

1. (a) 26–28 % **(1)**

   (b) The ammonia is cooled **(1)** so that it liquefies (turns into a liquid) **(1)**.

2. (a) Iron is used as catalyst **(1)** to increase the rate of reaction **(1)**.

   (b) Only some of the nitrogen and hydrogen react/ammonia breaks down again to form nitrogen and hydrogen **(1)**, so recycling stops them being wasted/gives them another chance to react **(1)**.

3. (a) $N_2 + 3H_2 \rightleftharpoons 2NH_3$ **(1)**

   (b) The temperature could be reduced **(1)** and the pressure could be increased **(1)**.

42. Equilibrium

1. (a) The yield increases **(1)**.

   (b) The rate will increase **(1)** because the reacting particles will collide more often **(1)**.

2. (a) The rates of the forward and backward reactions become exactly the same **(1)** in a closed system **(1)**.

   (b) There are fewer molecules of gas on the right (of the equation) **(1)**, so increasing the pressure moves the equilibrium to the right **(1)**, but very high pressures are expensive (or dangerous) **(1)**, so an optimum pressure of about 200 atmospheres is chosen **(1)**.

43. Alcohols

1. Ethanol reacts with sodium to produce hydrogen (second box ticked) **(1)**.

2. (a) Its name ends in 'ol' **(1)**; it contains an –OH group **(1)**.

   (b) fuel/solvent **(1)**

3. $2C_3H_7OH + 9O_2 \rightarrow 6CO_2 + 8H_2O$ (**1** mark for **6**, **1** mark for **8**)

44. Carboxylic acids

1. Ethanoic acid reacts with carbonates to produce carbon dioxide (second box ticked) **(1)**.
2. (a) Its name ends in 'anoic acid' **(1)**; it contains a —COOH group **(1)**.
   (b) ester **(1)**
3. Ethanoic acid is a weak acid **(1)** so it is not ionised completely when dissolved in water **(1)**, but hydrochloric acid is a strong acid **(1)**, which does ionise completely **(1)**.

45. Esters

1. food flavourings **(1)**
2. (a) —COO– **(1)**
   (b) ethyl ethanoate **(1)**
   (c) to act as a catalyst **(1)** to speed up the reaction **(1)**
3. Esters have distinctive smells **(1)**, and they are volatile **(1)**, so they carry the smell in the air **(1)**.
4. Methyl butanoate contains a –COO– group **(1)**, and its name contains 'yl' and 'anoate' **(1)**.

46. Using organic chemicals

1. Any one advantage, for example: esters are cheap to make **(1)**; it is easier to make esters than to extract natural flavours **(1)**; a lot of artificial flavouring may be made **(1)**. Any one disadvantage, for example: only a few different substances are used so the flavour might not be natural **(1)**; some people do not like eating artificial substances **(1)**. Justified opinion: I think artificial food flavours are fine to use because they make the food cheaper **(1)**; I don't think artificial food flavours should be used because it means that the food is not natural **(1)**.
2. Two advantages for one mark each, for example: ethanol is a renewable resource **(1)**; it produces less carbon dioxide overall when burned than fuels such as petrol **(1)**. Two disadvantages for one mark each, for example: ethanol is very flammable **(1)**; the crop plants used to make the ethanol could have been used to feed people instead **(1)**.

47. Chemistry six mark question 3

Answers can be found on page 92.

Physics answers

48. X-rays

1. The metal hip shows up more clearly than the bone **(1)** because they both absorb X-rays but the metal absorbs better than the bone **(1)**.
2. Scientists found that exposure to X-rays increased the chance of getting cancer **(1)**. The benefits of checking teeth outweigh the health risks **(1)**, but this is not so with buying shoes **(1)**. *The answer should recognise the hazard of X-rays to health and compare the risks with the benefits of X-ray photographs in dental treatment and shoe buying.*

49. Ultrasound

1. (a) B **(1)**; C **(1)**
   (b) D **(1)**
2. (a) There is a reflection of the sound where it enters the stomach **(1)** and another when it reaches the opposite side of the stomach **(1)**.
   (b) time difference = 0.000 21 s − 0.000 05 s = 0.000 16 s **(1)**; distance travelled = 1500 m/s × 0.000 16 s **(1)** = 0.24 m **(1)**; so diameter of stomach = ½ × 0.24 m = 0.12 m **(1)**

50. Medical physics

1. (a) CT scan **(1)**
   (b) ultrasound **(1)**
   (c) X-ray photograph **(1)**
   (d) ultrasound (scan) **(1)**
2. (a) CT scans can give a detailed picture **(1)** of the inside of the body **(1)**.
   (b) Ultrasound does not damage health so can be used as often as necessary/can be used on unborn babies **(1)**. CT scans carry a risk of cancer **(1)** but ultrasound scans are less detailed than CT scans **(1)**, so CT scans are better when detail is required **(1)**.

51. Refraction in lenses

1. The lens refracts the sunlight to a point **(1)** so there is enough energy to burn the paper **(1)**. *It is actually the infrared light emitted by the Sun that causes the paper to burn. Infrared light is refracted in the same way as visible light.*
2. $\sin r = \dfrac{\sin 30°}{2.42}$ **(1)** = 0.5/2.42 = 0.2066 **(1)**, so r = 11.9° **(1)**

52. Images and ray diagrams

1. real **(1)**; smaller/diminished **(1)**

53. Real images in lenses

1. magnification = 6 cm/8 cm **(1)** = 0.75 **(1)**
2. Completed ray diagram. 1 mark for each ray accurately drawn **(3)**, 1 mark for correct positioning of image **(1)**. *Image should be same size and the same distance from the lens as the object.*

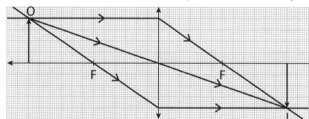

54. Virtual images in lenses

1. The image in a concave lens is always virtual **(1)** because the lens causes the refracted rays to diverge **(1)**.
2. Magnification = height of image/height of object = 2 **(1)**

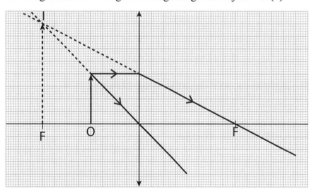

2 marks for drawing in two or three rays accurately. 1 mark for correctly marking in image.

55. The eye

1. cornea **(1)**, lens **(1)**
2. (a) iris **(1)** *The pupil is just a hole, and the iris controls the size of the hole.*
   (b) ciliary muscles and lens **(1)**
   (c) retina **(1)**

56. Range of vision

1. The ciliary muscles contract **(1)**, causing the lens to become fatter/more curved **(1)**, which bends the light more to form a sharp image on the retina **(1)**.

57. Correction of sight problems

1. **(a)** short-sighted **(1)**
   (b) eyeball longer than normal/lens more curved than normal **(1)**
   (c) Diverging/concave lenses should be used **(1)** as these will reduce the amount by which light entering the eye from distant objects converges **(1)**.
2. The lens in the eye is not bending the light enough **(1)** to produce a sharp image on the retina of a near object, so a converging/convex **(1)** lens is needed to assist in the converging of the light. *This happens because the ciliary muscles of older or tired eyes are unable to contract sufficiently, or in old eyes the lens becomes harder and unable to change shape.*

58. Power of a lens

1. B has a higher refractive index than C so bends the light more/has the shorter focal length **(1)** and is more curved than A so bends the light more **(1)**.
2. Lens Y bends the light more than lens X because it has a higher power **(1)**. As the lenses are the same shape, the difference must be caused by the refractive index of the material **(1)**, so Y is made of a material with the higher refractive index **(1)**.

59. Total internal reflection

1. glass, air **(1)**; air, glass **(1)**
2. $\sin c = 1/\text{refractive index} = 1/1.5$ **(1)** $= 0.67$ **(1)** $c = \sin^{-1} 0.67 = 42°$ **(1)**

60. Other uses of light

1. The endoscope contains optical fibres **(1)**. Light travels down the optical fibre into the stomach **(1)**. It is reflected by the walls of the stomach and travels back up the fibre to form an image **(1)**. *The image can be on a computer screen or it can be viewed through an eyepiece.*
2. Answers should provide one advantage, one disadvantage and a conclusion. Any one advantage from: no longer have to wear spectacles or contact lenses **(1)**; can focus at all distances without assistance **(1)**. Any one disadvantage from: there is always a risk of infection during surgery **(1)**; it is expensive **(1)**. Any conclusion from: either it is not worth the risk **(1)**; expense just to do away with spectacles/contact lenses **(1)**; OR the surgery is worth the risks/costs as it restores perfect sight **(1)**.

61. Physics six mark question 1

Answers can be found on page 92–3.

62. Centre of mass

1.

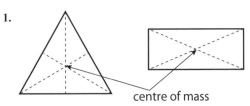

centre of mass

1 mark for correct lines on each object **(2)** and 1 mark for each centre of mass labelled **(2)**.

2. He cannot fix the letters C, L and D with a single nail **(1)** because their centre of mass is not on the wooden letter/is in the space around the letter **(1)**.

63. The pendulum

1. **(a)** $f = 30$ swings/60 seconds **(1)** $= 0.5$ **(1)** Hz
   (b) $T = 1/0.5$ Hz **(1)** $= 2$ **(1)** s
2. The period of the swing stays the same **(1)** because the period does not depend on the mass **(1)**, only the length **(1)**.

64. Turning effect and levers

1. **(a)** moment $= 8$ N $\times 0.12$ m **(1)** $= 0.96$ **(1)** Nm anticlockwise **(1)**
   (b) 0.96 Nm **(1)** *If the spanner is not moving the moments must be balanced, and so there must be a clockwise moment of the same size.*
2. Increase the (perpendicular) distance from the pivot to the point where the effort acts **(1)**; decrease the (perpendicular) distance between the pivot and the line of action of the load/move the load closer to the pivot **(1)**.

65. Moments and balance

1. If L is the distance of the child from the mid-point of the seesaw, anticlockwise moment $= 160$ N $\times 1.5$ m $= 240$ Nm, clockwise moment $= 200$ N $\times L$ m $= 200 L$ Nm **(1)**. When the see-saw is balanced $200 L$ Nm $= 240$ Nm **(1)**; $L = 240/200 = 1.2$ m **(1)**.

66. Stability

1. A **(1)**, as it has a wider base than B **(1)** and a lower centre of mass than C **(1)**.
2. For even a small movement away from the vertical the line of action of the weight of the bicycle and rider will fall outside the width of the tyres **(1)**, producing a moment that makes the bicycle topple over **(1)**.

67. Hydraulics

1. **(a)** $P = 8000$ N$/0.04$ m^2 **(1)** $= 200\ 000$ **(1)** Pa
   (b) The man can raise the weight of the car because he produces the pressure in the liquid **(1)** by applying a smaller force to the piston with the smaller cross-sectional area **(1)**.
2. pressure at brakes $= 160$ N$/0.002$ m^2 $= 80\ 000$ Pa **(1)**; pressure on foot-brake $= 80\ 000$ Pa; so force on foot brake $= P \times a = 80\ 000$ Pa $\times 0.0001$ m^2 **(1)** $= 8$ N **(1)**

68. Circular motion

1. **(a)** centripetal force **(1)**
   (b) an arrow pointing towards the centre of the circle **(1)**
   (c) The centripetal force is greater for the adult **(1)** because the adult has a larger mass **(1)**.
2. The frictional force between the tyre and road provides the centripetal force that makes the bicycle turn **(1)**. When the road is icy the frictional force is lower **(1)**, so the speed of the bicycle must be lower if it is to (reduce the centripetal force needed to) make the bicycle turn **(1)**.

69. Physics six mark question 2

Answers can be found on page 93.

70. Electromagnets

1. A permanent magnet is always magnetic **(1)**, so it could pick things up but not put them down **(1)**, but an electromagnet stops being magnetic when the current is turned off **(1)**.
2. When the 1st switch is closed a small current flows through the coil, making it an electromagnet **(1)**. This attracts the iron strip/the armature and closes the switch in the circuit with the larger current **(1)**.

71. The motor effect

1. (a) up **(1)**   (b) down **(1)**   (c) down **(1)**
2. Answers should give one similarity and one difference. Similarities: Both motors contain a coil and a magnet/the coil experiences a force when a current flows through the coil **(1)**. Differences: in the locomotive the current will be greater, the magnetic field stronger **(1)**.

72. Electromagnetic induction

1. A battery produces a direct current (d.c.) **(1)** so the magnetic field that is produced in the primary does not change **(1)**. Therefore no potential difference is induced in the secondary coil **(1)**.
2. The potential difference is induced due to the changing magnetic field **(1)** across the secondary coil produced by the alternating potential difference **(1)** in the primary coil.

73. Step-up and step-down transformers

1. The number of coils in the secondary is 4× the number on the primary **(1)** so the potential difference in the secondary is 4× the potential difference in the primary **(1)**; OR the ratio of the potential differences **(1)** is the same as the ratio of the number of coils **(1)**; OR $V_s = V_p \times n_s/n_p = 12 \times 200/50$ **(1)** = 48 **(1)** V.
2. Assuming 100% efficiency **(1)** power in primary = load = 3000 W, so $I_p = P/V_p$ = 3000 W/30 000 V **(1)** = 0.1 A **(1)**.

74. Switch mode transformers

1. Any three answers from: switch mode transformers are lighter/smaller **(1)**; do not have a large/heavy iron core **(1)**; operate at a higher frequency/50–200 kHz **(1)**; switch off/waste less power when not under load **(1)**.
2. Old TVs used traditional transformers, which draw power whenever they are plugged into the mains **(1)**. Modern TVs use switch mode transformers **(1)**, which use very little power when there is no load.

75. Physics six mark question 3

Answers can be found on page 93.

Six mark question answers

A basic answer is usually badly organised, has only basic information in it, does not use scientific words and includes poor spelling, punctuation and grammar.

A good answer usually contains accurate information and shows a clear understanding of the subject. The answer will have some structure and the candidate will have tried to use some scientific words, but it might not always be accurate and there may not be all the detail needed to answer the question. There will be a few errors with spelling, punctuation and grammar.

An excellent answer contains accurate information, is detailed and is supported by relevant examples. The answer will be well organised and will contain lots of relevant scientific words that are used in the correct way. The spelling, punctuation and grammar will be almost faultless.

10. Biology six mark question 1

A basic answer will include a simple explanation of one relevant fact, such as that oxygen is absorbed in the lungs *or* oxygen is carried round the body in blood.

A good answer will include a detailed description of how alveoli are adapted for diffusion *or* how red blood cells are adapted for carrying oxygen.

An excellent answer will include a clear, balanced and detailed explanation of the ventilation of lungs, the diffusion of oxygen gas across the alveoli into the blood, the combining of oxygen with haemoglobin in red blood cells and the release of oxygen in capillaries within organs.

Examples of points made in the response:
- Contraction and relaxation of muscles in the ribs and around the diaphragm cause ventilation/moving of air into and out of the lungs.
- The lungs have millions of alveoli which increase the surface area for diffusion.
- The alveoli are well supplied with blood capillaries to aid diffusion.
- The thin walls of capillaries and alveoli mean oxygen only has to diffuse a short distance between them.
- In the alveoli oxygen moves down its concentration gradient into red blood cells.
- In red blood cells the oxygen combines with haemoglobin to form oxyhaemoglobin.
- The blood circulation carries the blood around the body, reaching all the organs.
- In an organ, the oxyhaemoglobin breaks down to haemoglobin, releasing the oxygen to the tissues for respiration.

15. Biology six mark question 2

A basic answer may contain one relevant fact, such as that insulin causes cells to take glucose from the blood.

A good answer will include a detailed description of how insulin is secreted by the pancreas into the blood as blood glucose concentration increases, *or* why an insulin dose needs to consider the glucose content of food and the amount of exercise planned.

An excellent answer will include a clear, balanced and detailed explanation of the natural control of blood glucose concentration as well as the way diabetics carefully calculate injections of insulin.

Examples of points made in the response:
- Blood glucose concentration rises after a meal as glucose is absorbed from digested food in the gut.
- Blood glucose concentration falls as cells take in glucose for respiration or for storage as glycogen.
- Normally, as blood glucose concentration rises, the pancreas detects the rise and responds by secreting insulin.
- Insulin is a hormone that causes cells to take up more glucose from the blood.
- A person with type 1 diabetes does not produce insulin so their blood glucose concentration can rise dangerously high after a meal.
- Insulin injections help a type 1 diabetic control blood glucose concentration.
- Different meals contain different amounts of glucose and so will increase blood glucose concentration to different amounts.
- Exercise means more glucose is taken from the blood for muscle cell respiration.
- So the dose of injected insulin must be carefully calculated to make sure that blood glucose concentration doesn't go too high or too low.

Glucagon may be mentioned in the answer but, as it is not specifically asked for in the question, it will only be marked as part of the discussion of the control of natural blood glucose concentration.

23. Biology six mark question 3

A basic answer will describe one relevant fact, such as that deforestation damages the environment.

A detailed answer will include information on the ways that deforestation damages the environment *or* how oil palm plantations could contribute to human food and energy supplies in the future.

An excellent answer will include a clear, balanced and detailed description of benefits and problems related to replanting large areas of tropical forest with plantations.

Examples of points made in the response:

Benefits:

- The human population is growing so we need more food.
- We also need more fuel to use in engines and to make electricity.
- Using biofuel can be better than using fossil fuels to produce energy because it doesn't release as much carbon dioxide into the air. *Deforestation releases less carbon dioxide than burning fossil fuels, but it could still contribute to global warming.*
- Carbon dioxide from fossil fuels is adding to global warming.
- A benefit of growing oil palms is that it provides more money for the country.

Problems:

- Deforestation decreases biodiversity by replacing the plants and trees of the forest with plantation trees.
- Many of the animals that used to live in the forest will no longer be able to live and may become extinct.
- A problem with growing oil palms is that most of the money from the products goes to large companies, not local people.
- Deforestation will also cause pollution that can damage the environment.

32. Chemistry six mark question 1

A basic answer will include descriptions of some of the properties of one of the groups, with one of the trends in a group identified correctly.

A good answer will include descriptions of some of the properties of *both* of the groups, with one of the trends from both groups compared.

An excellent answer will include a clear, balanced and detailed description of several properties of both groups, with two trends compared and contrasted.

Examples of points made in the response:

Group 1 properties:

- Low density/float on water.
- All metals.
- Produce ions with a +1 charge.
- React with water to form hydrogen.
- Form alkaline hydroxides.
- All their atoms have one electron in their outer shell.

Group 1 trends:

- Become more reactive going down the group.
- Melting points and boiling points decrease going down the group.
- All solids at room temperature.

Group 7 properties:

- All non-metals.
- Produce ions with a −1 charge.
- All their atoms have seven electrons in their outer shell.

Group 7 trends:

- Become less reactive going down the group.
- Melting points and boiling points increase going down the group.
- Chlorine is a gas at room temperature; bromine is a liquid; iodine is a solid.

40. Chemistry six mark question 2

A basic answer will include a brief description of an environmental, social or economic impact of hard water. Little information from the table is used.

A good answer will include a clear description of a few environmental, social and economic impacts of hard water. Information from the table is used in support.

An excellent answer will include clear, balanced and detailed descriptions of environmental, social and economic impacts of hard water, fully supported using relevant knowledge and information from the table.

Examples of points made in the response:

Environmental impact:

- More energy use.
- Specified increase in energy use, e.g. from the table.
- More emissions from increased use of fuels.
- Increased use of limited resources, for example oil, gas or coal.
- More soap used.

Social impact:

- Hard water is good for the development and maintenance of bones and teeth.
- Hard water helps to reduce heart disease.
- Unpleasant appearance of soap scum or limescale.

Economic impact:

- Increased energy costs.
- Specified increase in energy cost, e.g. from the table.
- Increased cost from the need to use water softeners.
- Replacement or repair cost of damaged heating systems or kettles.
- Increased cost from extra soap use.

47. Chemistry six mark question 3

A basic answer will include a brief description of the titration, which may also include a basic risk assessment.

A good answer will include a good description of the titration, including a basic risk assessment, but with some details missing.

An excellent answer will include a clear, balanced and detailed description of the titration, with an appropriate risk assessment.

Examples of points made in the response:

Apparatus:

- burette
- pipette
- white tile
- conical flask

Substances:

- hydrochloric acid
- diluted oven cleaner
- indicator (need not be named)

Method:

- Hydrochloric acid in the burette.
- Known volume of diluted oven cleaner in the conical flask (using the pipette).
- Take burette reading at start and end.
- Add acid until end point or indicator colour change.
- Swirl to mix.
- Repeat experiment.

Risk assessment:

- Hydrochloric acid is irritant or corrosive.
- Oven cleaner is irritant or corrosive.
- Wear eye protection.
- Wear gloves.
- Use a pipette filler for the pipette.

61. Physics six mark question 1

A basic answer will include a brief description of a use or of the properties of X-rays and ultrasound, or a more detailed discussion of the use and properties of either X-rays or ultrasound.

A good answer will include a description of the properties of X-rays and ultrasound and relate them to their uses in imaging and therapy.

An excellent answer will include a clear, detailed and balanced description of the similarities and differences between X-rays and ultrasound. It will relate their properties to their use in imaging and therapy.

Examples of points made in the response:

Properties:

- X-rays are a form of electromagnetic radiation with a very high frequency/short wavelength.
- Ultrasound is high-frequency sound waves.
- X-rays pass through soft tissues but are absorbed by bone.
- Ultrasound is reflected off boundaries between tissues and organs.
- X-rays are a form of ionising radiation.
- X-rays are harmful/can cause cancer.
- Ultrasound is (thought to be) harmless.

Uses:

- X-rays form images on photographic film and/or on CCDs.
- CT scans build up a picture using X-rays from many angles.
- X-rays used to form images of bones/teeth/cancers.
- Ultrasound used to form images of soft tissues or foetus.
- Ultrasound images are formed by measuring the time taken for reflected waves to return to the sensor.

Similarities:

- Both used to form images of tissues and organs within the body.
- Both can be used in some form of medical treatment.

Differences:

- X-rays are harmful so must be used sparingly, workers must take precautions.
- Ultrasound can be used many times and no special precautions are needed.

69. Physics six mark question 2

A basic answer will include a brief discussion of the force acting on the seats on the roundabout.

A good answer will include a description of the centripetal force and one factor that affects it.

An excellent answer will include a clear and detailed explanation of centripetal force and the factors that affect it.

Examples of points made in the response:

- When the roundabout is rotating the velocity of each seat is changing.
- This is because its direction is constantly changing.
- The seats are therefore accelerating.
- A centripetal force is acting on each seat towards the centre of the roundabout.
- If a chain breaks the seat will move in a straight line.

- The chain will snap if the centripetal force exceeds its breaking force.
- Increasing the speed of the roundabout increases the centripetal force.
- The greater the mass of the seat and rider, the greater the centripetal force.
- The smaller the radius of the circle that the seats move in, the greater the centripetal force.

75. Physics six mark question 3

A basic answer will include a brief explanation of how a transformer works.

A good answer will include an explanation of how a transformer functions as well as its structure, and may include some calculations.

An excellent answer will include a clear, detailed and balanced explanation of the function of the particular transformer, including calculations of the effect of the transformer on the current.

Examples of points made in the response:

- Mains electricity is harmful/can cause a serious electric shock.
- The model railway uses a lower/safer voltage.
- This transformer is a step-down transformer.
- The output potential difference is lower than the input potential difference.
- A transformer consists of a two coils of wire wound onto a (soft) iron core.
- The coil connected to the input is the primary coil.
- The coil connected to the output is the secondary coil.
- When an alternating current flows in the primary coil there is a changing magnetic field.
- This induces an alternating current in the secondary coil.
- There are a larger number of turns of wire on the primary coil than the secondary coil.
- The ratio of potential difference in this example is $V_p/V_s = 230/11.5 = 20/1$
- So the ratio of turns of the transformer coils n_p/n_s (primary to secondary) is 20/1 and the primary coil has 20 times more turns than the secondary.
- The transformer transfers electrical power from the primary to the secondary.
- The maximum power transferred to the output = 11.5 V × 3 A = 34.5 W.
- If the transformer is 100% efficient the input power is 34.5 W.
- And the current in the primary coil is 11.5 V × 3 A/230 V = 0.15 A.

Published by Pearson Education Limited, Edinburgh Gate, Harlow, Essex, CM20 2JE.

www.pearsonschoolsandfecolleges.co.uk

Copies of official specifications for all AQA qualifications may be found on the AQA website: www.aqa.org.uk

Text and original illustrations © Pearson Education Limited 2013
Edited by Judith Head and Florence Production Ltd
Typeset and illustrated by Tech-Set Ltd, Gateshead
Cover illustration by Miriam Sturdee

The rights of Peter Ellis, Sue Kearsey and Nigel Saunders to be identified as authors of this work have been asserted by them in accordance with the Copyright, Designs and Patents Act 1988.

First published 2013

17 16 15 14 13
10 9 8 7 6 5 4 3 2 1

British Library Cataloguing in Publication Data
A catalogue record for this book is available from the British Library

ISBN 978 1 447 94249 8

Printed in Slovakia by Neografia

Acknowledgements
The authors and publisher would like to thank the following individuals and organisations for permission to reproduce copyright material:

Tables
Table from Tuomisto, H.L., 2010. "Food Security and Protein Supply - Cultured meat a solution", Aspects of Applied Biology 102, pp.99-104, www.academia.edu/722467/Food_Security_and-protein_supply_*cultured_meat_a_solution, copyright © Hanna Tuomisto; American Society of Agronomy for a table from "Postanthesis Moderate Wetting Drying Improves Both Quality and Quantity of Rice Yield" by Hao Zhang, Shenfeng Zhang, Jianchang Yang, et al, Agronomy Journal, ACSESS-Alliance of Crop, Soil, and Environmental Science Societies, 1 May 2008, copyright © 2008 by the American Society of Agronomy; and Table adapted from data about bluebell flowers, www.naturescalendar.org.uk. Reproduced with permission.
All other images © Pearson Education

Every effort has been made to contact copyright holders of material reproduced in this book. Any omissions will be rectified in subsequent printings if notice is given to the publishers.

In the writing of this book, no AQA examiners authored sections relevant to examination papers for which they have responsibility.